Powering Progress Solar Energy Solutions for Commercial Buildings

Giovanni

TABLE OF CONTENTS

Chapter 3: Assessing Solar Feasibility for Commercial Buildings

Chapter 4: Designing a Solar System for Commercial Buildings

Chapter 9: Conclusion: Embracing Solar Energy for Commercial Buildings 136

Chapter 1: Introduction to Solar Energy

The Importance of Renewable Energy

In today's rapidly changing world, the importance of renewable energy, particularly solar energy, cannot be overstated. As concerns about climate change, air pollution, and dwindling fossil fuel reserves continue to grow, the adoption of renewable energy sources has become a global imperative. This subchapter aims to shed light on the significance of renewable energy, with a particular focus on solar power, as a solution for powering commercial buildings.

First and foremost, renewable energy is clean and sustainable. Unlike traditional sources such as coal and oil, solar energy produces no greenhouse gas emissions or harmful pollutants when converted into electricity. By harnessing the power of the sun, commercial buildings can significantly reduce their carbon footprint, contributing to a healthier and more sustainable environment for everyone. Moreover, solar energy is an abundant resource that will never run out, ensuring a long-term and reliable energy supply.

From an economic perspective, the adoption of solar energy can bring substantial benefits. Commercial buildings that utilize solar power can drastically reduce their dependence on expensive fossil fuels, resulting in significant cost savings over time. As the cost of solar technology continues to decline, the return on investment for solar installations has become even more appealing. Moreover, solar energy systems require minimal maintenance, offering long-term financial stability and predictability in energy costs for commercial building owners.

Furthermore, solar energy promotes energy independence and resilience. By generating electricity on-site, commercial buildings can become less reliant on the grid, reducing the risk of power outages and disruptions. This is particularly crucial during times of natural disasters or emergencies, when reliable access to electricity is vital.

The importance of renewable energy, specifically solar power, extends beyond environmental and economic considerations. It also presents an opportunity to foster innovation and create jobs. The growing demand for solar installations has led to the emergence of a thriving industry, creating employment opportunities and driving economic growth. By investing in solar energy solutions for commercial buildings, we can stimulate technological advancements and contribute to a more sustainable and prosperous future.

In conclusion, the importance of renewable energy, particularly solar power, cannot be overstated. It offers a clean, sustainable, and abundant source of electricity, reducing greenhouse gas emissions, improving air quality, and conserving fossil fuel reserves. Additionally, solar energy provides economic benefits, such as cost savings, energy independence, and job creation. By embracing solar energy solutions for commercial buildings, we can power progress towards a more sustainable and resilient future for everyone.

Global Energy Crisis

In recent years, the world has been facing an unprecedented challenge - the global energy crisis. This crisis, driven by the ever-increasing demand for energy and the finite nature of traditional fossil fuel resources, has forced us to rethink our approach to power generation. As we seek sustainable alternatives, solar energy has emerged as a frontrunner, offering a plethora of solutions for commercial buildings.

The global energy crisis is a pressing issue that affects everyone, from individuals to businesses and governments. Our reliance on fossil fuels has not only led to environmental degradation but has also put a strain on the global economy. With energy demands projected to double by 2050, the need for a clean, renewable, and abundant energy source has never been more urgent.

Solar energy provides a viable solution to this crisis. Unlike fossil fuels, solar power is renewable, harnessing the sun's energy and converting it into electricity without emitting greenhouse gases or pollutants. This not only mitigates the environmental impact but also reduces our dependence on costly and depleting fossil fuels.

Commercial buildings play a pivotal role in the global energy consumption landscape. They consume a significant amount of energy for heating, cooling, lighting, and other power-intensive operations. By adopting solar energy solutions, these buildings can reduce their carbon footprint and contribute to a more sustainable future.

Solar energy solutions for commercial buildings come in various forms. Solar panels, also known as photovoltaic (PV) systems, can be installed on rooftops or integrated into building facades, harnessing sunlight and converting it into electricity. This renewable energy source can power a building's electrical systems, reducing or even eliminating the reliance on traditional power grids.

Additionally, solar thermal systems can generate heat for water heating or space heating, further reducing the energy demand from conventional sources. These solutions not only provide long-term cost savings but also offer a reliable and independent energy supply, safeguarding against price fluctuations and disruptions in the energy market.

Adopting solar energy solutions is not just environmentally responsible but also financially beneficial. Governments and organizations worldwide are providing incentives and grants to promote solar energy adoption, making it an attractive investment for businesses. By reducing energy costs and achieving energy independence, commercial buildings can improve their long-term financial outlook while contributing to a cleaner and more sustainable future.

In conclusion, the global energy crisis necessitates a shift towards renewable energy sources, and solar energy stands at the forefront of this transition. With its abundance, environmental benefits, and financial advantages, solar energy solutions have the potential to power progress and offer a sustainable future for commercial buildings and the world at large. It is time for everyone to embrace solar energy as a solution to the global energy crisis.

Environmental Concerns

In recent years, the global community has become increasingly aware of the urgent need to address environmental concerns. The impact of human activities on the planet has led to the depletion of natural resources, the destruction of ecosystems, and the alarming rise in greenhouse gas emissions. As a result, there is a growing consensus that we must take immediate action to mitigate the effects of climate change and ensure a sustainable future for generations to come.

One of the most pressing environmental concerns is the overreliance on fossil fuels for energy production. Traditional sources such as coal, oil, and natural gas not only contribute significantly to greenhouse gas emissions but also pose severe health risks due to air pollution. The extraction and burning of these non-renewable resources have led to devastating consequences, including habitat destruction, water pollution, and the disruption of fragile ecosystems.

In light of these challenges, the adoption of renewable energy sources, such as solar energy, has emerged as a crucial solution. Solar energy harnesses the power of the sun to generate clean electricity without any harmful emissions or environmental degradation. It offers an abundant, sustainable, and virtually limitless source of energy, making it an ideal alternative to fossil fuels.

Solar energy solutions for commercial buildings hold immense potential for addressing environmental concerns. By integrating solar panels into the infrastructure of commercial buildings, we can significantly reduce greenhouse gas emissions and curb our reliance on non-renewable energy sources. This transition to solar energy not only helps combat climate change but also offers numerous economic benefits, such as reduced energy bills and long-term cost savings.

Moreover, the deployment of solar energy systems in commercial buildings can have a transformative effect on the surrounding community. It serves as a visible demonstration of a commitment to sustainability and environmental stewardship, inspiring others to follow suit. The integration of solar energy solutions can also contribute to local job creation and foster the growth of a green economy.

In conclusion, addressing environmental concerns is a shared responsibility that requires collective action from all individuals and industries. Solar energy solutions for commercial buildings offer a viable and sustainable pathway towards a greener future. By transitioning to clean and renewable energy sources, we can mitigate the effects of climate change, protect the environment, and create a more prosperous and sustainable world for everyone.

Economic Advantages

In today's ever-changing world, the need for sustainable and renewable energy sources has become increasingly important. Solar energy has emerged as one of the most viable and promising options to meet the growing energy demands while reducing our carbon footprint. In this subchapter, we will explore the economic advantages of solar energy and how it can benefit commercial buildings and businesses.

First and foremost, solar energy offers significant cost savings. By harnessing the power of the sun, businesses can reduce their reliance on traditional energy sources, such as fossil fuels, which tend to be expensive due to fluctuating prices and limited availability. Solar energy, on the other hand, provides a stable and predictable energy source, allowing businesses to better forecast and manage their energy costs. Additionally, with advancements in technology and economies of scale, the cost of solar panels and installation has significantly decreased, making it more affordable for commercial buildings to adopt solar energy systems.

Another economic advantage of solar energy is the potential for long-term financial gains. Many governments and local authorities offer attractive incentives and tax breaks to businesses that invest in renewable energy, including solar power. These incentives can come in the form of grants, rebates, or even feed-in tariffs, where excess energy generated by solar panels can be sold back to the grid. By taking advantage of these incentives, businesses can not only offset the initial costs of installing solar energy systems but also generate additional revenue streams over time.

Furthermore, solar energy can enhance a business's brand and reputation. In today's environmentally conscious society, consumers are increasingly favoring companies that demonstrate a commitment to sustainability and reducing their carbon footprint. By going solar, businesses can position themselves as environmentally responsible and attract a broader customer base, leading to increased customer loyalty and revenue growth.

Lastly, solar energy offers businesses long-term energy security. As traditional energy sources become scarcer and more expensive, solar power provides a renewable and abundant energy source that businesses can rely on for decades to come. By investing in solar energy systems, commercial buildings can secure their energy supply and reduce the risks associated with energy price volatility.

In conclusion, solar energy provides numerous economic advantages for commercial buildings and businesses. From cost savings and financial incentives to enhanced brand reputation and long-term energy security, solar power offers a sustainable and economically viable solution. Embracing solar energy not only benefits businesses' bottom line but also contributes to a greener and more sustainable future for all.

Solar Energy Basics

Introduction:
In this subchapter, we will explore the fundamentals of solar energy, providing you with a comprehensive understanding of this renewable energy source. Whether you are new to solar energy or already have some knowledge on the topic, this section will serve as a valuable resource for everyone interested in harnessing the power of the sun.

What is Solar Energy?
Solar energy refers to the energy derived from the sun's radiation, which can be converted into usable electricity or heat. This clean and abundant source of renewable energy has gained significant attention in recent years due to its potential to combat climate change and reduce dependence on fossil fuels.

How Does Solar Energy Work?
Solar energy is harnessed through the use of photovoltaic (PV) cells, which convert sunlight into electricity. These cells are typically made from semiconductor materials, such as silicon, that absorb photons from sunlight and release electrons, creating an electric current. This direct current (DC) is then converted into alternating current (AC) by an inverter, making it suitable for powering commercial buildings.

Benefits of Solar Energy:
Solar energy offers numerous benefits, making it an attractive solution for commercial buildings. Firstly, it is a renewable and sustainable energy source, meaning it will never run out as long as the sun continues to shine. Additionally, solar energy produces no greenhouse gas emissions during operation, helping to mitigate the impacts of climate change. Moreover, installing solar panels can significantly

reduce electricity bills, providing substantial long-term savings for commercial building owners.

Integration into Commercial Buildings: Solar energy systems can be seamlessly integrated into commercial buildings, providing a reliable and clean source of power. These systems consist of solar panels, mounting structures, inverters, and electrical components. When properly designed and installed, solar panels can generate a substantial amount of electricity to meet a building's energy needs.

Conclusion:
Solar energy is a promising solution for commercial buildings, offering a sustainable, cost-effective, and environmentally friendly alternative to traditional energy sources. By understanding the basics of solar energy, you can make informed decisions about incorporating solar power into your commercial building, contributing to a greener future for all.

How Solar Energy Works

Solar energy is a powerful, renewable resource that has the potential to revolutionize the way we power our world. In this subchapter, we will delve into the fascinating world of solar energy and explore how it works to provide clean and sustainable power.

At its core, solar energy is harnessed from the sun's rays using photovoltaic (PV) cells. These cells are typically made of silicon, a semiconductor material that enables the conversion of sunlight into electricity. When sunlight strikes the PV cells, it excites the electrons, causing them to flow and generate a direct current (DC). This DC electricity is then converted into alternating current (AC) through an inverter, making it compatible with the electrical grid or for use in commercial buildings.

The sun's energy is abundant, and by capturing even a fraction of it, we can generate significant amounts of electricity. However, solar panels need to be strategically positioned to maximize their efficiency. This involves considering factors such as the tilt and orientation of the panels, as well as any potential shading from nearby structures or trees.

Once installed, solar panels require minimal maintenance, making them a cost-effective solution for commercial buildings. Routine cleaning and inspections ensure optimal performance, but overall, solar energy systems are known for their durability and longevity.

Solar energy is not just limited to generating electricity. It can also be harnessed for heating purposes through solar thermal systems. These systems use sunlight to heat water or other fluids, which can then be used for space heating, hot water supply, or even industrial processes.

Solar thermal systems are particularly popular in commercial buildings that require large volumes of hot water.

In addition to its environmental benefits, solar energy also offers financial advantages. Many countries and governments offer incentives, such as tax credits and rebates, to encourage businesses and individuals to adopt solar energy. These incentives can significantly reduce the upfront costs and provide a quick return on investment.

By harnessing the power of the sun, solar energy can play a crucial role in reducing carbon emissions and combating climate change. It offers a sustainable and reliable source of power, making it an ideal choice for commercial buildings looking to transition to clean energy solutions.

In conclusion, solar energy works by converting sunlight into electricity through photovoltaic cells. It is a renewable resource that requires minimal maintenance and offers both environmental and financial benefits. By embracing solar energy, commercial buildings can power their progress towards a more sustainable future.

Types of Solar Panels

Solar energy is a rapidly growing industry that offers a sustainable and clean energy solution for commercial buildings. One of the key components of solar energy systems is solar panels, which convert sunlight into electricity. In this subchapter, we will explore the different types of solar panels available in the market today.

1. Monocrystalline Solar Panels: Monocrystalline panels are the most efficient and expensive type of solar panels. They are made from a single crystal structure, typically silicon, which allows for higher energy conversion rates. Monocrystalline panels are known for their sleek black appearance and excellent performance in low-light conditions.

2. Polycrystalline Solar Panels: Polycrystalline panels are made from multiple silicon crystals, resulting in a lower energy conversion efficiency compared to monocrystalline panels. However, they are more cost-effective and widely available. Polycrystalline panels have a blue hue and are a popular choice for commercial buildings due to their affordability.

3. Thin-Film Solar Panels: Thin-film panels are made from a variety of materials, including amorphous silicon, cadmium telluride, and copper indium gallium selenide. These panels are lightweight, flexible, and can be seamlessly integrated into various surfaces, such as building facades and windows. Although thin-film panels have lower energy conversion efficiency compared to crystalline panels, they perform better in high-temperature environments.

4. Bifacial Solar Panels: Bifacial panels have the ability to absorb sunlight from both sides, increasing their energy production. These

panels can capture reflected and diffused light, making them a suitable choice for installations in areas with high albedo, such as snow-covered landscapes or buildings with reflective surfaces.

5. Building-Integrated Photovoltaics (BIPV): BIPV systems integrate solar panels directly into the building envelope, such as roofing materials, windows, or facades. These panels blend seamlessly with the building's architecture, providing both energy generation and aesthetic appeal. BIPV systems are gaining popularity in commercial buildings, as they offer enhanced design possibilities and improved energy efficiency.

When considering solar energy solutions for commercial buildings, it is important to assess the specific requirements and goals of the project. The choice of solar panels will depend on factors such as budget, available space, energy consumption, and aesthetic preferences. Consulting with solar energy professionals can provide valuable insights and help determine the most suitable type of solar panels for a commercial building.

In summary, the types of solar panels available range from monocrystalline and polycrystalline panels, offering high efficiency at a higher cost, to thin-film panels that provide flexibility and integration options. Bifacial panels and BIPV systems offer additional advantages in specific scenarios. With continuous advancements in solar technology, the solar energy industry is well-positioned to power progress in commercial buildings, providing sustainable and clean energy solutions for a better future.

Solar Energy Efficiency

In this subchapter, we will explore the concept of solar energy efficiency and its significance in our journey towards a sustainable future. As the demand for clean energy solutions continues to rise, solar power has emerged as a frontrunner in the race towards a greener planet. Solar energy not only reduces our dependence on fossil fuels but also provides an abundant and renewable source of power for various applications. However, to fully harness the potential of solar energy, it is imperative to understand the importance of solar energy efficiency.

Solar energy efficiency refers to the ability of solar panels and systems to convert sunlight into usable energy effectively. By maximizing this efficiency, we can optimize the power output of solar panels, making them more cost-effective and environmentally friendly. Improving solar energy efficiency is crucial for both individual users and large-scale commercial buildings, as it directly impacts the overall performance and financial viability of solar energy systems.

One of the key factors influencing solar energy efficiency is the quality and design of solar panels. Advances in technology have led to the development of more efficient solar panels with higher conversion rates. These panels are designed to absorb a broader spectrum of sunlight, including both direct and indirect sunlight, thereby increasing their overall efficiency. Additionally, innovations in materials and manufacturing processes have made solar panels more durable and resistant to external factors such as temperature fluctuations and shading.

Furthermore, regular maintenance and monitoring of solar energy systems play a pivotal role in maintaining and enhancing their efficiency. Cleaning the solar panels to remove dust, dirt, and debris can significantly improve their performance. Additionally, monitoring the energy output and identifying any deviations or issues promptly can help rectify them before they impact the overall efficiency of the system.

Investing in energy storage solutions, such as batteries, is another strategy to enhance solar energy efficiency. By storing surplus energy generated during peak sunlight hours, users can ensure a consistent and reliable power supply, even when the sun is not shining. This not only increases the usability of solar energy but also reduces the reliance on grid power during non-sunny periods.

In conclusion, solar energy efficiency is a crucial aspect of harnessing the true potential of solar power. By improving the design and quality of solar panels, conducting regular maintenance, and investing in energy storage solutions, we can maximize the efficiency and effectiveness of solar energy systems. Embracing solar energy efficiency is not only beneficial for the environment but also for individuals and commercial buildings, as it offers a sustainable and cost-effective solution for meeting our energy needs. Let us all join hands in the pursuit of a brighter and greener future through solar energy efficiency.

Chapter 2: Benefits of Solar Energy for Commercial Buildings

Cost Savings

In today's world, where climate change and rising energy costs are major concerns, finding ways to save on expenses while promoting sustainable practices is crucial. This subchapter explores the cost savings associated with solar energy solutions for commercial buildings, highlighting the numerous benefits that can be achieved by harnessing the power of the sun.

Solar energy offers a wide range of financial advantages, making it an attractive option for businesses of all sizes. One of the primary benefits is the reduction in electricity bills. By installing solar panels on commercial buildings, companies can generate their own electricity and reduce their dependence on the grid. This leads to significant cost savings over time, as the sun's energy is free and abundant. In fact, studies have shown that businesses can save up to 75% on their electricity bills by adopting solar energy solutions.

Additionally, solar energy systems require minimal maintenance and have a long lifespan. With proper care, solar panels can last for more than 25 years, providing a reliable source of energy for decades. This longevity translates into long-term savings for businesses, as they can avoid the costs associated with frequent repairs or replacements.

Another cost-saving aspect of solar energy is the potential for generating revenue through net metering. Net metering allows businesses to sell excess electricity generated by their solar panels back to the grid. This not only offsets the initial investment but also creates an additional income stream. By taking advantage of net metering programs, businesses can turn their solar energy systems into profitable assets.

Furthermore, adopting solar energy solutions can lead to tax incentives and rebates. Many governments and local authorities offer financial incentives to encourage businesses to go solar. These incentives can include tax credits, grants, or accelerated depreciation, which can significantly reduce the upfront costs of installing solar panels.

Lastly, investing in solar energy solutions can enhance a company's brand image and attract environmentally conscious customers. As sustainability becomes increasingly important to consumers, businesses that demonstrate their commitment to clean energy can gain a competitive edge. This can result in increased customer loyalty and higher sales, ultimately contributing to the company's bottom line.

In conclusion, the cost savings associated with solar energy solutions for commercial buildings are abundant and diverse. From reduced electricity bills to revenue generation and tax incentives, solar energy provides a multitude of financial advantages for businesses. Additionally, the long lifespan and minimal maintenance requirements of solar panels make them a cost-effective investment. By embracing solar energy, businesses can not only save money but also contribute to a more sustainable future.

Reduced Electricity Expenses

In today's world, where rising electricity costs and environmental concerns are at the forefront, finding ways to reduce electricity expenses is crucial for individuals and businesses alike. One of the most effective solutions to this problem lies in harnessing the power of solar energy. Solar energy is a clean, renewable, and abundant source of power that can significantly reduce your electricity bills while also benefiting the environment.

Solar energy systems for commercial buildings have gained immense popularity in recent years due to their numerous advantages. By installing solar panels on the rooftops or open spaces of your commercial building, you can generate your own electricity and reduce your dependence on the traditional power grid. As a result, your electricity expenses will be significantly reduced, providing you with long-term cost savings.

Not only do solar energy systems help in reducing electricity expenses, but they also provide a predictable and stable energy cost for businesses. Unlike traditional electricity prices that fluctuate due to market conditions, solar energy allows you to generate your own power at a fixed cost. This stability in energy costs can help businesses plan their budgets more effectively, ensuring financial stability and predictability.

Moreover, solar energy systems require minimal maintenance and have a long lifespan, making them a cost-effective investment in the long run. Once the initial installation costs are recovered, the electricity generated by solar panels is essentially free. Additionally, many governments and local authorities offer incentives and tax credits to businesses that invest in solar energy, further reducing the overall expenses.

Another significant advantage of solar energy systems is their positive impact on the environment. By generating clean energy from the sun, businesses can significantly reduce their carbon footprint and contribute to mitigating climate change. Solar energy produces no greenhouse gas emissions, air pollution, or noise pollution, making it a sustainable and eco-friendly choice for commercial buildings.

In conclusion, solar energy solutions offer a practical and sustainable way to reduce electricity expenses for commercial buildings. By harnessing the power of the sun, businesses can not only save money but also contribute to a greener future. With the availability of government incentives and tax credits, the time is ripe for every business to consider integrating solar energy solutions into their operations. Don't miss out on the opportunity to power progress and make a positive impact on both your finances and the environment.

Tax Incentives and Rebates

In the world of solar energy, tax incentives and rebates play a crucial role in driving the adoption of solar power systems. These incentives are designed to encourage individuals and businesses to invest in renewable energy technologies, such as solar panels, by offering financial benefits that help offset the initial costs.

For every one of us, whether we are homeowners, small business owners, or large corporations, tax incentives and rebates are available to make solar energy solutions more affordable and accessible. These incentives can come in various forms, such as federal tax credits, state rebates, and utility company incentives.

One of the most significant tax incentives available to everyone is the Federal Investment Tax Credit (ITC). This credit allows individuals and businesses to deduct a percentage of the cost of installing a solar power system from their federal taxes. Currently, the ITC offers a 26% credit for systems installed before the end of 2022, and a 22% credit for systems installed in 2023. This tax credit can significantly reduce the upfront costs of going solar, making it an attractive option for many.

Additionally, many states offer their own rebate programs to further incentivize solar energy adoption. These rebates vary by state but can provide significant financial assistance for solar installations. Some states also offer property tax exemptions, sales tax exemptions, or low-interest loans to support solar energy projects.

Furthermore, utility companies often provide incentives to their customers for installing solar power systems. These incentives can include cash rebates, net metering programs, and feed-in tariffs. Net metering allows solar system owners to sell excess electricity back to the utility grid, effectively reducing their monthly energy bills.

It is crucial for anyone interested in solar energy solutions to research and understand the available tax incentives and rebates specific to their location. Consulting with a solar energy professional or contacting local government agencies can provide valuable information on the incentives available in your area.

By taking advantage of these tax incentives and rebates, solar energy becomes a more financially viable option for everyone. Not only do these incentives make solar power systems more affordable, but they also contribute to a sustainable future by reducing greenhouse gas emissions and dependence on fossil fuels.

In conclusion, tax incentives and rebates are essential tools in promoting solar energy solutions for commercial buildings and beyond. With the support of federal and state governments, as well as utility companies, solar power becomes a viable and attractive option for individuals and businesses alike. By harnessing the power of the sun, we can contribute to a cleaner and more sustainable future for generations to come.

Long-Term Financial Benefits

When it comes to investing in solar energy solutions for commercial buildings, the long-term financial benefits cannot be overstated. Solar power is not only a sustainable and environmentally friendly energy source, but it can also significantly reduce operational costs and provide an attractive return on investment.

One of the most compelling financial advantages of solar energy is the potential for substantial energy savings. By harnessing the power of the sun, commercial buildings can generate their own electricity, reducing or even eliminating their reliance on traditional energy sources. This can lead to significant cost savings on monthly utility bills, especially as the price of electricity continues to rise.

Additionally, solar energy systems often come with generous government incentives and tax credits. These financial incentives can help offset the initial investment cost of installing a solar power system, making the transition to solar energy even more financially appealing. By taking advantage of these incentives, businesses can recoup their investment in a shorter period of time and start reaping the financial benefits sooner.

Another long-term financial benefit of solar energy is the potential for revenue generation. In some cases, commercial buildings with solar power systems can produce more electricity than they consume. This excess energy can be sold back to the grid, allowing businesses to earn additional income through net metering programs or feed-in tariffs. This additional revenue stream can further enhance the financial viability of solar energy solutions and provide a steady source of income for years to come.

Furthermore, installing solar energy systems can increase the value of commercial properties. Studies have shown that buildings equipped with solar panels tend to have higher resale values compared to those without. This is due to the growing demand for sustainable and energy-efficient buildings in the real estate market. Therefore, investing in solar energy solutions not only provides immediate financial benefits but also enhances the long-term value of commercial properties.

In conclusion, embracing solar energy solutions for commercial buildings offers a wide range of long-term financial benefits. From reducing energy costs and taking advantage of government incentives to generating additional revenue and increasing property value, solar power is a smart and financially rewarding choice for businesses of all sizes. By harnessing the power of the sun, commercial buildings can power progress towards a more sustainable and prosperous future.

Environmental Sustainability

In recent years, the world has witnessed an increasing concern for environmental sustainability. The need to find alternative sources of energy that are clean, renewable, and efficient has become more pressing than ever before. Solar energy has emerged as a viable solution to address the energy crisis and combat the adverse effects of climate change. This subchapter explores the significance of solar energy in promoting environmental sustainability and its potential to power commercial buildings.

Solar energy is derived from the sun's rays, making it an abundant and renewable resource. Unlike fossil fuels, solar energy does not produce harmful greenhouse gas emissions or contribute to air pollution. By harnessing the power of the sun, we can reduce our reliance on non-renewable energy sources and mitigate the negative impacts of conventional energy generation on the environment.

Commercial buildings are one of the largest consumers of energy, accounting for a significant portion of the global energy consumption. Adopting solar energy solutions in these buildings can lead to substantial environmental benefits. Solar panels can be installed on rooftops, facades, or open spaces surrounding the building, converting sunlight into electricity. This renewable energy source can power various operations within the building, including lighting, heating, and cooling systems. By using solar energy, commercial buildings can reduce their carbon footprint, decrease energy costs, and contribute to a greener future.

Solar energy not only provides environmental advantages but also offers long-term economic benefits. With advancements in technology, the cost of solar panels has significantly decreased, making solar energy more affordable and accessible to a wide range of commercial buildings. Moreover, solar energy systems require minimal maintenance and have a lifespan of up to 25 years. This translates into long-term savings on energy bills and a reliable source of electricity for years to come.

In addition to its environmental and economic benefits, solar energy also enhances the reputation of commercial buildings as environmentally responsible entities. By embracing solar energy solutions, these buildings demonstrate their commitment to sustainability, attracting environmentally conscious tenants, customers, and investors.

Solar energy is a powerful tool for promoting environmental sustainability in commercial buildings. By harnessing the sun's energy, we can reduce greenhouse gas emissions, decrease energy costs, and create a greener future for generations to come. Embracing solar energy solutions is not only an investment in the environment but also a step towards a sustainable and resilient future.

Reduced Carbon Footprint

As the world grapples with the effects of climate change, it has become increasingly imperative to find sustainable and environmentally friendly solutions to our energy needs. Solar energy has emerged as a key player in this quest, offering immense potential to reduce our carbon footprint and contribute to a greener future.

In this subchapter, we will delve into the concept of reducing carbon footprint through solar energy solutions in commercial buildings. This information is relevant to everyone, whether you are an individual concerned about the environment or a business owner looking to cut down on energy costs and enhance your corporate social responsibility.

Solar energy is a renewable and clean source of power that harnesses the sun's abundant energy to generate electricity. By utilizing solar panels, commercial buildings can significantly reduce their reliance on fossil fuels, which are major contributors to greenhouse gas emissions. By transitioning to solar energy, businesses can play a pivotal role in mitigating climate change and protecting the planet for future generations.

The benefits of reducing carbon footprint through solar energy are manifold. Firstly, solar power offers a sustainable and inexhaustible energy source, ensuring that businesses have a reliable and long-term solution for their energy needs. This allows for greater energy independence and stability, as well as protection against volatile energy prices.

Furthermore, solar energy systems require minimal maintenance and have a long lifespan, making them cost-effective in the long run. By investing in solar power, businesses can not only save money on their energy bills but also benefit from various financial incentives offered by governments and utility companies. These incentives, such as tax credits and rebates, make solar energy an attractive option for businesses of all sizes.

Additionally, adopting solar energy solutions can enhance a company's reputation and brand image. In today's environmentally conscious world, consumers and stakeholders are increasingly demanding sustainable practices from businesses. By reducing their carbon footprint through solar energy, companies can showcase their commitment to sustainability and attract environmentally conscious customers.

In conclusion, reducing carbon footprint through solar energy solutions in commercial buildings is a crucial step towards a sustainable future. Whether you are an individual or a business owner, embracing solar power offers numerous benefits, including cost savings, energy independence, and a positive environmental impact. By harnessing the power of the sun, we can power progress and pave the way for a greener and brighter tomorrow.

Lowered Greenhouse Gas Emissions

In today's world, the issue of climate change and its adverse effects on the environment has become a major concern for everyone. The burning of fossil fuels for energy generation is one of the leading contributors to greenhouse gas emissions, which are responsible for global warming. As a result, there has been a growing need to find sustainable and renewable energy sources to power our progress towards a greener future. Solar energy has emerged as a key solution in this quest for lowering greenhouse gas emissions.

Solar energy is derived from the sun's radiation and is harnessed through solar panels or photovoltaic cells. Unlike traditional energy generation methods, solar power does not release harmful pollutants or greenhouse gases into the atmosphere. By utilizing solar energy, we can significantly reduce our carbon footprint and mitigate the impact of climate change.

One of the primary benefits of solar energy is its ability to generate electricity without emitting any greenhouse gases. This is achieved by converting sunlight into electricity through a clean and renewable process. By installing solar panels on commercial buildings, businesses can directly generate electricity on-site, reducing their reliance on fossil fuel-powered grids. This not only helps in lowering greenhouse gas emissions but also provides a reliable and cost-effective energy source for businesses.

Furthermore, solar energy can be integrated with energy storage systems, allowing excess electricity generated during the day to be stored and used during peak demand hours or at night. This promotes grid stability and reduces the need for backup power from fossil fuel-based sources, which further contributes to lowering greenhouse gas emissions.

Another significant advantage of solar energy is its scalability. Solar panels can be installed on rooftops, parking lots, or even as standalone solar farms. This flexibility allows solar energy to be utilized in various commercial settings, regardless of the available space. By embracing solar energy solutions, commercial buildings can take a proactive approach towards reducing their carbon footprint and demonstrating their commitment to sustainability.

In conclusion, solar energy offers a viable solution for lowering greenhouse gas emissions in commercial buildings. Its ability to generate electricity without releasing harmful pollutants makes it an ideal choice for businesses looking to transition towards a greener future. By harnessing solar power, we can contribute to the global efforts in combating climate change, ensuring a sustainable and environmentally friendly path for progress.

Positive Brand Image

In today's highly competitive business landscape, building a positive brand image is essential for companies looking to thrive in the solar energy industry. A brand's image represents its reputation and the way it is perceived by customers, stakeholders, and the general public. It encompasses the company's values, mission, and the overall experience it offers. The development of a strong and positive brand image is paramount, as it not only attracts and retains customers but also creates a sense of trust and credibility in the market.

For solar energy companies, a positive brand image is especially crucial. As the world becomes more environmentally conscious, the demand for sustainable energy solutions has skyrocketed. Customers are not only looking for efficient and cost-effective solar systems but also want to align themselves with brands that share their values. By building a positive brand image, solar companies can differentiate themselves from competitors and establish themselves as trustworthy and reliable partners in the renewable energy transition.

To create a positive brand image, solar energy companies need to focus on several key aspects. Firstly, they should emphasize their commitment to sustainability and environmental responsibility. By showcasing initiatives such as reducing their carbon footprint, using eco-friendly materials, and supporting community projects, companies can demonstrate their dedication to a greener future.

Secondly, solar energy companies should prioritize customer satisfaction. By providing exceptional products and services, addressing customer concerns promptly, and offering competitive prices, companies can build a loyal customer base and establish a positive reputation in the market.

Furthermore, solar companies should also engage in effective communication and public relations strategies. By sharing success stories, case studies, and educational content about solar energy, companies can position themselves as thought leaders in the industry. This will not only enhance their brand image but also help in educating the public about the benefits of solar energy.

Lastly, solar energy companies should actively engage with their stakeholders, including employees, investors, and local communities. By involving these groups in decision-making processes and demonstrating a commitment to social responsibility, companies can create a positive brand image that resonates with a wide range of audiences.

In conclusion, building a positive brand image is crucial for solar energy companies to thrive in today's competitive market. By focusing on sustainability, customer satisfaction, communication, and stakeholder engagement, companies can establish themselves as leaders in the industry and attract a diverse range of customers. A positive brand image not only helps in acquiring new customers but also fosters long-term relationships and creates a sense of trust and credibility in the solar energy market.

Chapter 3: Assessing Solar Feasibility for Commercial Buildings

Solar Potential Assessment

Assessing the solar potential of a commercial building is a crucial step in harnessing the power of solar energy. Understanding the solar potential of a building helps in determining the viability and feasibility of installing solar panels, and ultimately, maximizing the benefits of solar energy. In this subchapter, we will delve into the various aspects of solar potential assessment and how it plays a pivotal role in powering progress towards sustainable commercial buildings.

Solar potential assessment begins with a thorough analysis of the building's location. Factors such as latitude, longitude, and local climate patterns are taken into consideration to determine the amount of solar radiation the building receives throughout the year. This assessment provides valuable insights into the solar energy potential of the site, allowing building owners and stakeholders to make informed decisions regarding solar energy adoption.

Another crucial aspect of solar potential assessment involves evaluating the building's orientation and shading. The orientation of a building affects its exposure to sunlight, with south-facing roofs generally receiving the maximum solar radiation. By analyzing the building's orientation and identifying potential shading obstacles such as nearby trees or tall buildings, experts can accurately estimate the solar potential and suggest appropriate measures to mitigate any shading issues.

Furthermore, an assessment of the building's roof space is essential. The size, shape, and structural integrity of the roof determine the feasibility of installing solar panels. A thorough examination of the roof's condition ensures that it can support the weight of solar panels and that any necessary repairs or reinforcements are made before installation.

Moreover, the assessment also involves evaluating the building's energy consumption patterns. By analyzing historical energy data and identifying peak usage periods, experts can determine the optimal system size and design to meet the building's energy needs. This enables building owners to make informed decisions regarding the installation of solar panels and energy storage systems, thus maximizing their energy savings and reducing reliance on the grid.

In conclusion, the solar potential assessment is a critical step towards harnessing the power of solar energy in commercial buildings. By taking into account various factors such as location, orientation, shading, roof space, and energy consumption patterns, building owners and stakeholders can make informed decisions regarding solar energy adoption. This subchapter aims to provide valuable insights and guidance to individuals and businesses interested in leveraging solar energy solutions, ultimately contributing to a sustainable and energy-efficient future for all.

Site Evaluation

Choosing the right site for a solar energy installation is crucial for maximizing its efficiency and effectiveness. Site evaluation involves assessing various factors to determine the suitability of a location for solar energy solutions. In this subchapter, we will explore the key aspects of site evaluation and how they impact the success of solar energy projects.

One of the primary factors to consider during site evaluation is the amount of sunlight available. Solar panels require direct sunlight to generate electricity, so it is essential to assess the amount of solar irradiance a site receives throughout the year. Factors such as shading from nearby buildings, trees, or other obstructions can significantly impact the solar potential of a location. By analyzing historical solar radiation data and conducting on-site shade studies, solar energy professionals can accurately determine the optimal placement of solar panels.

Another crucial aspect of site evaluation is the structural integrity of the building or land where the solar installation will take place. Commercial buildings should undergo a thorough structural assessment to ensure they can support the weight of solar panels and other equipment. Additionally, the orientation and tilt of the roof or available land should be considered to maximize the capture of solar energy. Evaluating the structural capacity and orientation of a site helps determine the feasibility and cost-effectiveness of a solar energy solution.

Environmental factors also play a significant role in site evaluation. The proximity to power lines, substations, or electrical infrastructure is crucial to assess the ease of grid connectivity. Additionally, the availability of suitable land or roof space for solar panel installation should be considered. Evaluating the environmental impact of a solar energy project is essential to ensure compliance with local regulations and minimize any adverse effects on the surrounding ecosystem.

In conclusion, site evaluation is a critical step in the process of implementing solar energy solutions for commercial buildings. By assessing factors such as solar irradiance, structural integrity, and environmental impact, solar energy professionals can determine the suitability of a location for solar installations. Careful site evaluation helps identify the most efficient and cost-effective solutions, maximizing the benefits of solar energy for commercial buildings.

Shading Analysis

When it comes to harnessing the power of solar energy, understanding shading analysis is crucial. Whether you are an individual homeowner looking to install solar panels or a commercial building owner interested in transitioning to renewable energy, shading analysis plays a significant role in maximizing the efficiency and effectiveness of your solar energy system.

Shading analysis refers to the assessment of how shading affects the performance of solar panels or any other solar energy system. Shading can occur due to various factors, such as nearby trees, buildings, or even temporary obstacles like debris or construction equipment. These shading obstructions can significantly impact the amount of sunlight that reaches the solar panels, thereby reducing their energy production potential.

To conduct a shading analysis, professionals use advanced tools and software to model and predict the shading patterns throughout the day and year. This analysis helps identify potential problem areas and allows for the optimization of solar panel placement to avoid or mitigate shading issues. By strategically positioning solar panels to minimize shading, you can maximize solar energy production and ensure a higher return on investment.

For homeowners, shading analysis can determine the best location on the property to install solar panels, taking into account the varying shade patterns throughout the year. This analysis can help homeowners avoid costly mistakes, such as installing panels in shaded areas that will significantly reduce energy production. By choosing the optimal location, homeowners can generate more electricity, reduce their reliance on the grid, and potentially even sell excess energy back to the utility company.

Similarly, for commercial buildings, shading analysis is essential to determine the most suitable areas for solar panel installation. By identifying potential shading obstacles, such as neighboring buildings or trees, businesses can optimize their solar energy systems and maximize their energy production. This not only helps to reduce operational costs but also demonstrates a commitment to sustainability and renewable energy, which can enhance a company's reputation and attract environmentally conscious customers.

In conclusion, shading analysis is a critical component of solar energy system design and installation. By understanding and mitigating shading issues, both homeowners and commercial building owners can harness the full potential of solar energy, save on electricity costs, and contribute to a cleaner and more sustainable future.

Solar Resource Mapping

Solar resource mapping plays a crucial role in maximizing the potential of solar energy and ensuring its efficient utilization in commercial buildings. By accurately assessing the solar resource, businesses can make informed decisions about the viability and design of solar energy systems. This subchapter explores the importance of solar resource mapping and its benefits for commercial buildings.

Solar resource mapping involves the measurement and analysis of solar irradiance, which is the amount of solar energy received on a given surface area. This data is collected using advanced technologies such as ground-based measurement stations, satellite imagery, and weather models. By mapping the solar resource, businesses can determine the optimal locations for solar panel installations and estimate their energy output.

One of the key benefits of solar resource mapping is its ability to assess the long-term solar potential of a specific location. By analyzing historical solar data, businesses can predict the solar resource availability throughout the year and identify potential variations. This information is crucial for designing solar energy systems that can consistently generate electricity and meet the energy demands of commercial buildings.

Furthermore, solar resource mapping allows businesses to accurately estimate the financial viability of solar energy projects. By assessing the solar resource and considering factors such as installation costs, energy prices, and available incentives, businesses can determine the return on investment and make informed decisions about adopting solar energy solutions. This helps businesses in optimizing their energy costs and reducing their reliance on fossil fuels.

Solar resource mapping also aids in the identification of potential shading and obstruction issues that may affect solar panel performance. By analyzing the topography and surrounding infrastructure, businesses can mitigate these issues through proper system design and placement. This ensures the maximum utilization of solar energy and minimizes any potential losses due to shading.

In conclusion, solar resource mapping is an essential tool for harnessing the potential of solar energy in commercial buildings. By accurately assessing the solar resource, businesses can make informed decisions about solar system design, estimate energy output, and determine the financial viability of solar energy projects. This not only helps in reducing energy costs but also contributes to a cleaner and more sustainable future for all.

Financial Evaluation

In today's fast-paced world, solar energy has emerged as a game-changer in the quest for sustainable and clean energy solutions. As the demand for renewable energy sources continues to increase, solar energy has become a popular choice for commercial buildings. However, before making the switch to solar power, it is crucial to conduct a thorough financial evaluation to assess the feasibility and benefits of this investment.

The financial evaluation of solar energy solutions is a comprehensive analysis that takes into account various factors. One of the key considerations is the initial cost of installing solar panels and related equipment. While the upfront investment may seem significant, it is important to remember that solar energy systems have long-term benefits that can offset these costs.

One of the primary advantages of solar energy is the potential for substantial savings on electricity bills. By harnessing the power of the sun, commercial buildings can significantly reduce their reliance on traditional energy sources. Over time, this can result in significant cost savings, making solar energy a financially attractive option.

In addition to lower energy costs, solar energy solutions often come with government incentives and tax credits. These financial incentives can further offset the initial investment and enhance the overall financial viability of the project. It is crucial to thoroughly research and understand the available incentives in your region to maximize the financial benefits of solar energy.

Furthermore, solar energy systems have a long lifespan, typically lasting for 25 to 30 years with minimal maintenance costs. This longevity ensures a reliable and consistent source of energy, reducing the risk of unexpected energy price hikes or disruptions. By investing in solar energy, commercial buildings can protect themselves from the volatility of the energy market and achieve long-term stability.

Moreover, integrating solar energy solutions can enhance the value of commercial buildings. As sustainability becomes an increasingly important factor for businesses and investors, solar-powered buildings are seen as more desirable and environmentally friendly. This can lead to higher property values, increased rental rates, and improved brand reputation.

In conclusion, conducting a thorough financial evaluation is crucial before making the switch to solar energy solutions for commercial buildings. By considering factors such as initial costs, potential savings, government incentives, and long-term benefits, businesses can make informed decisions that align with their financial goals. Solar energy offers significant advantages in terms of cost savings, energy independence, and enhanced property value. It is a powerful and sustainable solution that can power progress for commercial buildings and contribute to a greener future for everyone.

Calculating Energy Consumption

Understanding how to calculate energy consumption is crucial for anyone interested in solar energy solutions for commercial buildings. By accurately assessing the energy needs of a building, one can determine the optimal solar system size and design, leading to maximum energy savings and return on investment.

To calculate energy consumption, we must first understand the basic unit of energy measurement: the kilowatt-hour (kWh). A kilowatt-hour represents the amount of energy consumed by using one kilowatt of power for one hour. This unit is commonly used to measure electricity consumption.

The first step in calculating energy consumption is to gather data on the building's energy usage. This can be obtained through utility bills, which typically provide monthly or yearly consumption figures in kWh. It is important to gather data for a representative period to account for seasonal variations in energy usage.

Once the consumption data is collected, it is essential to analyze it to identify trends and patterns. This analysis can help identify peak demand periods, daily energy usage patterns, and potential areas for energy efficiency improvements. Energy audits can also be conducted to assess the energy performance of specific systems or equipment within the building.

Next, it is crucial to consider factors that may affect energy consumption, such as climate conditions, occupancy patterns, and the building's usage. For example, a building located in a hot climate may consume more energy for cooling purposes, while a building with high occupancy levels may have higher energy demands.

With the gathered data and analysis in hand, we can proceed to calculate the building's average daily energy consumption. This is done by summing up the total energy consumption over a specific period (e.g., a year) and dividing it by the number of days in that period. This average daily consumption figure will serve as a baseline for designing a solar energy system that can meet the building's energy needs.

By accurately calculating energy consumption, commercial building owners and solar energy professionals can ensure the proper sizing and design of solar systems. This allows for the maximum utilization of solar energy, leading to significant cost savings, reduced reliance on non-renewable energy sources, and a more sustainable future.

Whether you are a building owner, energy consultant, or simply interested in solar energy, understanding how to calculate energy consumption is a fundamental step in harnessing the power of the sun to power progress.

Estimating Solar System Costs

One of the primary considerations when it comes to adopting solar energy solutions for commercial buildings is accurately estimating the costs involved. Understanding the expenses associated with solar system installations is crucial for businesses and individuals looking to harness the power of the sun and make a positive impact on the environment.

The cost of implementing a solar system can vary depending on various factors, including the size of the building, the energy needs, and the location. It is essential to conduct a thorough assessment of the site to determine the optimal system size and design that suits the specific requirements. This assessment will help estimate the upfront costs and potential savings over the long term.

When estimating solar system costs, it is important to consider several key components. The first is the cost of solar panels, which make up a significant portion of the overall expenses. The type and efficiency of the panels will impact the upfront costs, but higher-quality panels often result in greater long-term savings.

Another crucial aspect to consider is the cost of inverters and other system components. Inverters convert the direct current (DC) generated by the solar panels into alternating current (AC) that can be used to power the building. These components play a vital role in ensuring the system operates efficiently and reliably.

Installation costs, including labor and permits, must also be taken into account. Hiring professionals who specialize in solar installations is highly recommended to ensure optimal system performance and compliance with local regulations. Additionally, factoring in ongoing maintenance costs is essential to ensure the system continues to operate at peak efficiency.

When estimating solar system costs, it is important to consider the financial incentives available. Governments and utility companies often provide incentives, such as tax credits and rebates, to encourage the adoption of solar energy. These incentives can significantly reduce the upfront costs and shorten the payback period for commercial buildings.

To accurately estimate solar system costs, it is advisable to consult with experienced solar energy professionals. They can provide detailed analyses and cost breakdowns specific to the building's requirements and location. By conducting a comprehensive assessment and considering all relevant factors, businesses and individuals can make informed decisions about adopting solar energy solutions and contribute to a sustainable future.

In conclusion, estimating solar system costs is a crucial step when considering solar energy solutions for commercial buildings. By thoroughly assessing the site, considering key components, factoring in installation and maintenance costs, and taking advantage of available incentives, businesses and individuals can make informed decisions and contribute to a greener future.

Return on Investment Analysis

In today's rapidly evolving world, the importance of sustainable energy sources cannot be overstated. As the demand for renewable energy grows, solar power has emerged as a viable solution for commercial buildings seeking to reduce their environmental impact while also improving their financial bottom line. This subchapter, titled "Return on Investment Analysis," will delve into the key aspects of calculating and understanding the return on investment (ROI) of solar energy solutions for commercial buildings.

Regardless of whether you are a business owner, an investor, or simply an individual interested in solar energy, understanding the financial implications of installing solar panels is crucial. By analyzing the ROI, you can assess the feasibility, profitability, and long-term benefits of utilizing solar energy in commercial buildings.

This subchapter will begin by explaining the concept of ROI and its significance in the solar energy context. It will explore the various components that contribute to ROI calculation, such as system cost, energy savings, incentives, tax credits, and maintenance expenses. Through detailed examples and case studies, readers will gain insights into how these factors influence the overall return on investment.

Moreover, this subchapter will highlight the potential risks and challenges associated with solar energy investments. It will discuss factors like fluctuating energy prices, technology obsolescence, and regulatory changes that can impact the ROI. By understanding these risks, readers will be better equipped to make informed decisions regarding solar energy solutions.

Additionally, this subchapter will emphasize the importance of considering the long-term benefits beyond immediate financial returns. It will touch upon the positive environmental impact of solar energy, including reduced carbon footprint and increased sustainability. Understanding the broader implications of solar energy adoption will enable individuals and businesses to align their goals with the greater good.

By the end of this subchapter, readers will have a comprehensive understanding of how to perform ROI analysis for solar energy solutions in commercial buildings. Armed with this knowledge, they will be able to evaluate the financial viability and potential benefits of solar energy investments.

Whether you are an entrepreneur looking to reduce operating costs, an investor seeking sustainable opportunities, or an individual interested in solar energy, this subchapter will equip you with the necessary tools to evaluate and appreciate the return on investment offered by solar energy solutions for commercial buildings.

Chapter 4: Designing a Solar System for Commercial Buildings

System Sizing and Design Considerations

When it comes to harnessing the power of solar energy for commercial buildings, system sizing and design considerations play a crucial role in optimizing efficiency and ensuring maximum benefits. This subchapter aims to provide a comprehensive overview of the factors that need to be taken into account while sizing and designing solar energy systems.

One of the primary considerations in system sizing is determining the energy demand of the building. This involves analyzing historical energy consumption patterns, understanding peak demand periods, and estimating future energy requirements. By accurately gauging the energy demand, the solar energy system can be sized accordingly to meet the building's needs.

Another important aspect is site assessment. Factors such as location, orientation, and available roof space need to be evaluated to determine the solar potential of the building. The angle and orientation of the solar panels play a vital role in optimizing energy generation. Additionally, shading analysis is crucial to identify any potential obstructions that may hinder solar panel performance.

Inverter selection is another critical consideration. The inverter converts the direct current (DC) energy generated by the solar panels into alternating current (AC) energy, which is used to power the building. Choosing the right inverter that matches the system size, voltage requirements, and expected loads is essential to ensure seamless energy conversion.

Battery storage is also an important consideration, especially for buildings that aim to achieve energy independence or require backup power during grid outages. Sizing the battery bank based on the building's energy requirements and usage patterns is crucial to ensure uninterrupted power supply.

Moreover, grid connection and net metering policies need to be taken into account during system design. Understanding the regulations and incentives provided by the utility company can help optimize energy savings and potentially generate revenue by exporting excess energy back to the grid.

Lastly, ongoing system monitoring and maintenance should also be considered. Implementing a robust monitoring system allows for real-time performance tracking, early detection of any issues, and proactive maintenance to ensure optimal system efficiency over time.

In conclusion, system sizing and design considerations are vital for the successful implementation of solar energy solutions in commercial buildings. By accurately assessing energy demand, conducting site evaluations, selecting appropriate inverters and batteries, understanding grid connection policies, and implementing monitoring systems, businesses can harness the full potential of solar energy to power their progress towards a sustainable future.

Determining Energy Needs

In the quest for sustainable and renewable energy sources, solar power has emerged as a beacon of hope. Its ability to harness the sun's energy and convert it into electricity has revolutionized the way we power our lives. As the world continues to grapple with the challenges of climate change and depleting fossil fuel reserves, solar energy has become a viable solution for commercial buildings.

One of the key steps in transitioning to solar energy is determining the energy needs of a commercial building. Understanding how much power a building requires is crucial in designing an effective solar energy system. By accurately assessing the energy needs, building owners can optimize the size and capacity of their solar panels, ensuring they generate enough electricity to meet the demand.

There are various factors to consider when determining energy needs. Firstly, the type of commercial building plays a significant role. Different buildings have different energy requirements based on their purpose and activities. For example, a manufacturing plant will consume more electricity than an office building. Therefore, it is essential to analyze the specific energy demands of the building based on its size, function, and equipment.

Another crucial aspect to consider is the historical energy consumption patterns of the building. By examining past utility bills, owners can identify trends and patterns in energy usage. This data can help in estimating future energy needs and identifying areas where energy efficiency improvements can be made.

Additionally, it is important to consider the geographical location of the building. The amount of sunlight a building receives throughout the year can vary depending on its position on the globe. By analyzing solar radiation data specific to the building's location, solar energy experts can determine the potential energy output of a solar panel system in that area.

Lastly, it is essential to factor in any future changes or expansions in the building. If there are plans to expand the facility or introduce new equipment, the energy needs will undoubtedly increase. Taking these future developments into account is crucial to ensure the solar energy system is scalable and can accommodate the growing demand.

Determining energy needs is a vital step in the journey towards solar energy adoption for commercial buildings. By accurately assessing the energy requirements based on building type, historical consumption patterns, location, and future plans, building owners can design and implement an efficient and effective solar energy system. This transition not only helps in reducing carbon emissions but also provides long-term cost savings and contributes to a cleaner and more sustainable future for everyone.

Selecting Solar Panel Types

When it comes to harnessing the power of solar energy, choosing the right solar panel type is essential. With advancements in technology and a wide range of options available in the market, it is crucial to understand the different types of solar panels and their suitability for various applications. In this subchapter, we will explore the various solar panel types and provide insights to help you make an informed decision.

1.	Monocrystalline	Solar	Panels: Monocrystalline panels are known for their high efficiency and sleek appearance. They are made from a single crystal structure, resulting in better performance in low-light conditions. Monocrystalline panels are an excellent choice for commercial buildings with limited roof space or those seeking maximum energy output.

2.	Polycrystalline	Solar	Panels: Polycrystalline panels are cost-effective and widely available. They are made by melting multiple fragments of silicon, resulting in a less uniform appearance. Although they have slightly lower efficiency compared to monocrystalline panels, they are still an excellent choice for commercial buildings with ample roof space.

3.	Thin-Film	Solar	Panels: Thin-film panels are lightweight and flexible, making them suitable for unconventional applications such as curved surfaces or portable solar systems. While they have lower efficiency compared to crystalline panels, they perform better in high temperatures and shading situations.

4.	Bifacial	Solar	Panels:
Bifacial panels can generate electricity from both sides, capturing sunlight from the front and reflecting light from the rear. This unique design increases the overall energy output, making them ideal for commercial buildings with reflective surfaces or elevated installations.

5.	Building-Integrated	Photovoltaics	(BIPV):
BIPV panels are specifically designed to replace traditional building materials such as windows, roof tiles, or facades. They seamlessly blend into the building's architecture, providing an aesthetically pleasing solution while generating clean energy.

When selecting solar panels, it is crucial to consider factors such as efficiency, available space, budget, and specific requirements of your commercial building. Consulting with a reputable solar energy provider can help you assess your needs and determine the most suitable panel type for your project.

Remember that solar panel technology is continually evolving, and it is essential to stay updated with the latest advancements. By selecting the most appropriate solar panel type, you can maximize your energy generation, reduce your carbon footprint, and contribute to a sustainable future.

Whether you are a business owner, building manager, or an individual interested in solar energy, understanding the various solar panel types is crucial for making an informed decision. Evaluate your needs, consult experts, and embark on your solar journey to power progress and contribute to a greener tomorrow.

Optimizing System Performance

In the rapidly evolving world of solar energy, optimizing system performance is crucial for maximizing the benefits of solar power in commercial buildings. As solar energy continues to gain popularity as a clean and sustainable alternative, it is important for everyone, from industry professionals to building owners, to understand how to make the most of this technology.

When it comes to solar energy systems, performance optimization involves various factors that can significantly impact the overall efficiency and effectiveness of the system. One key aspect is the proper design and installation of the solar panels. By ensuring that the panels are positioned correctly and receive maximum sunlight exposure throughout the day, their energy output can be maximized.

Regular maintenance and cleaning of the solar panels are also essential for optimal performance. Dust, debris, and other environmental factors can accumulate on the surface of the panels, reducing their efficiency. By regularly inspecting and cleaning the panels, the system's performance can be enhanced, leading to increased energy production.

Additionally, monitoring and analyzing the system's performance data is crucial for identifying any potential issues or areas of improvement. Utilizing advanced monitoring technologies, such as real-time data collection and analytics, can provide valuable insights into the system's performance, allowing for timely adjustments and optimizations.

Moreover, integrating energy storage solutions with solar energy systems can further enhance their performance. Energy storage systems can store excess energy generated by the solar panels, allowing for its use during periods of low sunlight or high energy demand. This not only maximizes the utilization of solar energy but also provides a reliable and uninterrupted power supply.

For building owners and operators, optimizing system performance can lead to significant financial benefits. By maximizing energy production and reducing reliance on conventional energy sources, solar energy systems can help lower electricity bills and decrease overall operational costs. Furthermore, the integration of solar energy solutions can also enhance a building's sustainability profile, attracting environmentally conscious consumers and investors.

In conclusion, optimizing system performance is crucial for harnessing the full potential of solar energy in commercial buildings. By ensuring proper design, installation, maintenance, and monitoring, solar energy systems can operate at their peak efficiency, providing sustainable and cost-effective energy solutions. Embracing solar energy and its optimization not only benefits the environment but also offers financial advantages, making it a win-win solution for everyone.

Integration with Existing Infrastructure

As solar energy continues to gain momentum as a sustainable and cost-effective solution for commercial buildings, the need for seamless integration with existing infrastructure becomes paramount. This subchapter explores the various aspects of integrating solar energy solutions into commercial buildings, addressing the concerns and challenges faced by individuals and businesses interested in harnessing the power of the sun.

One of the key advantages of solar energy is its compatibility with existing infrastructure. Whether you own a small business or manage a large commercial property, integrating solar panels into your building's structure can enhance your energy efficiency and reduce your reliance on traditional power sources. This integration can be achieved through a range of techniques, including rooftop installations, façade-integrated solar panels, and ground-mounted arrays.

Rooftop installations are a popular choice for many commercial buildings, as they utilize the available roof space without requiring any additional land. By mounting solar panels on the rooftop, businesses can generate electricity while utilizing space that would otherwise remain unused. Additionally, rooftop installations can help offset the energy consumption of the building, leading to substantial cost savings over time.

Façade-integrated solar panels offer an aesthetically pleasing solution for buildings with limited rooftop space. By incorporating solar panels into the building's exterior design, businesses can seamlessly blend sustainability with architectural appeal. These panels can be integrated into windows, cladding, or shading elements, allowing for the generation of clean energy without compromising the building's visual appeal.

For larger commercial properties with ample land available, ground-mounted arrays provide an excellent option for solar energy integration. These arrays consist of multiple solar panels mounted on the ground, either at a fixed tilt or with tracking systems to maximize energy production. Ground-mounted arrays are particularly beneficial for businesses with large energy demands, as they can be expanded to accommodate increased electricity needs.

To ensure a successful integration, it is essential to consider the compatibility of solar energy systems with existing electrical infrastructure. This may involve assessing the building's electrical capacity, upgrading wiring and distribution panels, and implementing appropriate safety measures. Working with experienced solar energy professionals can help ensure a seamless integration process, from design and installation to grid connection and ongoing maintenance.

By integrating solar energy solutions into commercial buildings, businesses can not only reduce their carbon footprint but also benefit from long-term cost savings and increased energy independence. With the flexibility offered by rooftop installations, façade-integrated panels, and ground-mounted arrays, there is a suitable solution for every commercial building. Embracing solar energy integration is a step towards a sustainable future, where renewable resources power progress for all.

Incorporating Solar Panels

Solar energy is rapidly gaining popularity as a clean and renewable source of power. It is being harnessed by individuals, businesses, and even governments to reduce their carbon footprint and contribute to a sustainable future. One of the most effective ways to utilize solar energy is by incorporating solar panels into commercial buildings. This subchapter explores the benefits and considerations of incorporating solar panels, providing valuable insights for anyone interested in solar energy solutions.

Solar panels, also known as photovoltaic (PV) panels, convert sunlight into electricity. By installing these panels on the rooftops or facades of commercial buildings, businesses can generate their own renewable energy and reduce their reliance on traditional power sources. The benefits of incorporating solar panels are multifaceted, encompassing environmental, economic, and social advantages.

From an environmental perspective, solar energy is a clean and sustainable alternative to fossil fuels. By using solar panels, commercial buildings can significantly reduce their greenhouse gas emissions, contributing to the fight against climate change. Solar energy also helps to conserve precious natural resources, as it does not require the extraction or burning of fossil fuels.

Economically, incorporating solar panels can lead to substantial cost savings in the long run. Once the initial investment is made, solar panels can generate electricity for decades with minimal maintenance costs. Businesses can also take advantage of various incentives and tax credits offered by governments to promote renewable energy adoption. Additionally, solar panels can act as a hedge against rising electricity prices, providing businesses with more control over their energy expenses.

Socially, incorporating solar panels demonstrates a commitment to sustainability and responsible business practices. It enhances a company's reputation and can attract environmentally conscious customers, investors, and partners. Moreover, solar energy creates job opportunities in the rapidly growing renewable energy sector, contributing to economic growth and job creation.

However, there are several considerations that businesses should keep in mind when incorporating solar panels. Factors such as location, shading, roof condition, and building orientation can impact the efficiency and effectiveness of solar panels. It is essential to conduct a thorough site assessment and consult with solar energy experts to determine the feasibility and optimal design for solar panel installation.

In conclusion, incorporating solar panels into commercial buildings offers numerous benefits for businesses and the environment. It enables businesses to generate their own clean and renewable energy, reducing their carbon footprint and contributing to a sustainable future. From cost savings to improved reputation, solar energy solutions are a win-win for businesses and society as a whole. By embracing solar power, we can power progress towards a cleaner and greener world.

Electrical System Integration

In the quest for a sustainable future, solar energy has emerged as a promising solution for commercial buildings. As the world recognizes the need to transition to renewable sources, harnessing the power of the sun has become increasingly important. One crucial aspect of solar energy solutions for commercial buildings is electrical system integration.

Electrical system integration refers to the process of seamlessly incorporating solar energy systems into the existing electrical infrastructure of a building. This integration allows for the efficient and effective utilization of solar power, maximizing its benefits while minimizing any disruptions to the building's operations.

Solar energy systems, such as photovoltaic (PV) panels, generate direct current (DC) electricity. However, most commercial buildings operate on alternating current (AC) power. Therefore, a key challenge in electrical system integration is the conversion of DC electricity into AC electricity. This is achieved through the use of inverters, which convert the DC power generated by solar panels into AC power that can be used by the building's electrical systems.

Another crucial aspect of electrical system integration is the connection between the solar energy system and the building's main electrical grid. This connection ensures that excess solar power can be fed back into the grid, allowing the building to earn credits or even generate income through net metering or feed-in tariff programs. On the other hand, when the solar system does not generate enough power, the building can draw electricity from the grid as needed.

To ensure a seamless integration, electrical system integration also involves the installation of monitoring and control systems. These systems enable real-time monitoring of the solar energy system's performance, allowing building owners to track energy production and consumption. By analyzing this data, adjustments can be made to optimize the system's efficiency and identify any potential issues.

Furthermore, electrical system integration also considers safety measures and compliance with relevant regulations. This includes proper grounding, protection against electrical faults, and adherence to local building codes and standards. It is crucial to ensure that the electrical system integration is carried out by qualified professionals to guarantee the safety and effectiveness of the solar energy solution.

In conclusion, electrical system integration is a vital component of solar energy solutions for commercial buildings. It enables the seamless incorporation of solar power into existing electrical infrastructure, maximizing energy efficiency and reducing environmental impact. By understanding the intricacies of electrical system integration, building owners can harness the full potential of solar energy and contribute to a greener and more sustainable future for all.

Roof and Structural Considerations

When it comes to harnessing the power of solar energy for commercial buildings, roof and structural considerations play a crucial role. The design and integrity of a building's roof and supporting structure can greatly impact the efficiency, functionality, and long-term viability of a solar energy system. In this subchapter, we will explore the key factors and considerations that must be taken into account when integrating solar energy solutions into commercial buildings.

First and foremost, the suitability of the roof for solar installations must be assessed. Factors such as the roof's orientation, tilt, and shading must be evaluated to determine the optimal placement and positioning of solar panels. South-facing roofs with minimal shading are generally the most ideal for maximizing solar energy capture. Additionally, the structural capacity of the roof should be assessed to ensure it can safely support the added weight of solar panels and any associated equipment.

Furthermore, the condition and age of the roof should be considered. If the roof is nearing the end of its lifespan, it may be wise to address any necessary repairs or replacement before installing solar panels. This will prevent the need for potentially costly and disruptive removal and reinstallation of the solar system in the future.

Another important consideration is the electrical infrastructure and wiring within the building. The solar energy system must be seamlessly integrated into the existing electrical system to ensure efficient energy transfer and distribution. This may require upgrades or modifications to the building's electrical infrastructure, which should be carefully planned and executed to ensure compliance with safety standards and regulations.

In addition to the roof, the structural integrity of the building itself must be evaluated. Solar energy systems can exert significant additional loads on the building's structure, especially in areas prone to high wind or snow loads. Engineering assessments should be conducted to determine the structural capacity and the need for any reinforcing measures.

Lastly, it is important to consider future expansion and flexibility. Commercial buildings may undergo changes in usage or require additional space for solar installations as energy needs grow. Therefore, the design and installation of the solar energy system should allow for easy scalability and adaptability to accommodate future expansion.

In conclusion, roof and structural considerations are essential when implementing solar energy solutions for commercial buildings. By carefully assessing the suitability of the roof, ensuring structural integrity, and planning for future flexibility, businesses can harness the full potential of solar energy while maximizing the return on investment.

Chapter 5: Installation and Maintenance of Solar Systems

Finding Qualified Solar Installers

When it comes to implementing solar energy solutions for commercial buildings, finding qualified solar installers is crucial. The success of your solar project heavily depends on the expertise and experience of the professionals you choose to work with. This subchapter will guide you on how to identify and select the right solar installers for your commercial building.

First and foremost, it is essential to understand the qualifications and certifications that solar installers should possess. Look for installers who are certified by recognized organizations such as the North American Board of Certified Energy Practitioners (NABCEP). NABCEP certification ensures that the installer has met rigorous standards and possesses the necessary skills to design and install solar energy systems.

Additionally, consider the experience of the solar installers. A long-standing track record indicates their ability to tackle complex projects successfully. Ask for references and check their past work to ensure they have experience in installing solar systems similar to what you require for your commercial building.

It is also important to assess the installer's knowledge of local regulations and permits. Each region may have specific requirements and permits for solar installations. A qualified installer should be well-versed in these regulations, ensuring compliance with local laws and minimizing any potential delays or issues during the installation process.

To further evaluate their expertise, ask the installers about their training and ongoing professional development. The solar industry is rapidly evolving, and staying updated with the latest technologies and best practices is crucial. Installers who invest in continuous education demonstrate their commitment to providing high-quality services.

When selecting solar installers, consider their reputation and customer reviews. Online platforms and local business directories can provide valuable insights into their past clients' experiences. Positive feedback and testimonials are indicators of their professionalism and reliability.

Lastly, don't forget to request detailed proposals and estimates from multiple installers. This will help you compare costs, system designs, warranties, and any additional services they may offer. It is crucial to choose an installer who can deliver a solution that meets your specific requirements while providing the best value for your investment.

In conclusion, finding qualified solar installers is vital for the successful implementation of solar energy solutions in commercial buildings. By considering their qualifications, experience, knowledge of local regulations, ongoing training, reputation, and pricing, you can make an informed decision and ensure a smooth and efficient solar installation process. Remember, investing in reputable and qualified solar installers will guarantee the long-term performance and benefits of your solar energy system.

Researching Local Solar Contractors

When it comes to harnessing the power of solar energy for commercial buildings, finding a reliable and experienced solar contractor is crucial. Whether you are looking to install solar panels on your rooftop or implement a complete solar energy solution, researching local solar contractors is an essential step towards powering progress. This subchapter aims to guide everyone interested in solar energy on how to find the right contractor for their specific needs.

One of the first steps in researching local solar contractors is to gather recommendations and references. Seek advice from friends, colleagues, or other businesses that have already undergone solar installations. Their experiences and insights can provide valuable information about the quality of work, customer service, and overall satisfaction with the contractor.

Another effective method is to explore online resources that list and rate solar contractors. Websites specializing in renewable energy or local business directories often include customer reviews and ratings. These platforms can help in narrowing down the options based on the contractor's reputation and track record.

Once you have compiled a list of potential contractors, it is important to verify their credentials and qualifications. Check if they are licensed and have the necessary certifications to install solar systems. Research their experience in the industry, looking for a track record of successful installations and satisfied clients.

The next step is to reach out to the shortlisted contractors and schedule initial consultations. During these meetings, discuss your project requirements, goals, and budget. A reliable contractor should be able to provide you with a detailed proposal, including cost estimates, system design, and estimated energy savings.

To further ensure the quality of work, ask for references from previous clients. Contact these references to inquire about their experience working with the contractor, the project's timeline, and any issues encountered. If possible, visit one or more of their completed projects to see the quality of their work firsthand.

Lastly, compare the proposals and estimates provided by different contractors. Evaluate not only the cost but also the quality of equipment, warranties, and maintenance plans offered. Remember that the cheapest option may not always be the best choice, as long-term reliability and efficiency are equally important.

By thoroughly researching local solar contractors, you can make an informed decision and find a reliable partner to turn your solar energy vision into reality. With the right contractor, your commercial building can become a shining example of sustainable progress and environmental responsibility.

Evaluating Installer Credentials

When it comes to solar energy solutions for commercial buildings, the expertise and credibility of the installer play a crucial role in ensuring the success and efficiency of the project. With the growing popularity of solar energy, it's becoming increasingly important for businesses and individuals to evaluate the credentials of installers before making any commitments. This subchapter aims to provide a comprehensive guide on how to assess installer credentials, enabling readers to make informed decisions.

First and foremost, one must consider the installer's certification and qualifications. Reputable installers should possess certifications such as the North American Board of Certified Energy Practitioners (NABCEP) or equivalent. These certifications validate the installer's knowledge and experience in solar energy systems. Additionally, evaluating the installer's level of experience is crucial. Ask for references and inquire about their previous projects, ensuring they have successfully completed installations similar to your requirements.

Another essential aspect to consider is the installer's licensing and insurance. A legitimate installer should hold the necessary licenses and permits required by local authorities. This ensures compliance with regulations and gives peace of mind that the installation is being carried out by professionals. Additionally, verifying that the installer has liability insurance protects you from any potential damages that may occur during the installation process.

Next, it's important to assess the installer's track record. Look for reviews and testimonials from previous clients to gain insights into their overall performance and customer satisfaction. Reputable installers should be transparent about their track record and readily provide references upon request.

Furthermore, evaluating the installer's warranties and after-sales support is crucial. A reliable installer should offer comprehensive warranties on both the equipment and installation workmanship. This ensures that any potential issues or repairs are covered, giving you peace of mind over the long-term performance of your solar energy system.

Lastly, consider the installer's commitment to ongoing education and staying updated with the latest industry advancements. The solar energy sector is rapidly evolving, and installers should continuously update their knowledge and skills to provide cutting-edge solutions.

In conclusion, evaluating the credentials of solar energy installers is vital for ensuring a successful and efficient installation. By considering factors such as certifications, experience, licensing, insurance, track record, warranties, and ongoing education, individuals and businesses can make informed decisions and choose the right installer for their commercial solar energy project.

Obtaining Multiple Quotes

In the world of solar energy, obtaining multiple quotes is a crucial step in making informed decisions and maximizing the benefits of solar power for commercial buildings. Whether you are a business owner, property manager, or simply an advocate for sustainable energy solutions, understanding the importance of obtaining multiple quotes can save you time, money, and potential headaches in the long run.

When it comes to solar energy systems, there are a plethora of options available in the market. Different manufacturers, installation companies, and financing options can make the process overwhelming. However, by obtaining multiple quotes, you can compare various aspects of each offer, enabling you to choose the most suitable solution for your specific needs.

One of the key reasons to obtain multiple quotes is to ensure that you are getting the best value for your investment. Solar energy systems can be a significant financial commitment, and it is essential to evaluate different quotes to find the most cost-effective solution. By comparing prices, warranties, and system specifications, you can identify any potential discrepancies or hidden costs, allowing you to make an informed decision.

Furthermore, obtaining multiple quotes enables you to assess the reputation and credibility of different solar energy companies. It is essential to work with experienced and reliable professionals who can deliver high-quality installations and provide ongoing support. By requesting quotes from multiple companies, you can review their track record, customer reviews, and certifications, ensuring that you choose a reputable partner for your solar energy project.

Additionally, obtaining multiple quotes gives you the opportunity to explore various financing options. Different companies may offer different financing plans, such as upfront payment, leasing, or power purchase agreements. By comparing the terms, interest rates, and payment structures of each quote, you can select the financing option that best aligns with your budget and long-term financial goals.

In conclusion, obtaining multiple quotes is a critical step in the process of powering progress with solar energy solutions for commercial buildings. By comparing prices, evaluating reputations, and exploring financing options, you can make an informed decision that maximizes the benefits of solar power for your specific needs. Remember, solar energy is a long-term investment, and obtaining multiple quotes ensures that you choose the right system and partner to embark on this sustainable journey.

Solar System Maintenance

Proper maintenance is crucial for the efficient functioning and longevity of any solar energy system. By ensuring regular maintenance, you can maximize the benefits of solar energy and minimize any potential issues that may arise. This subchapter aims to provide an overview of solar system maintenance, covering essential tips and best practices.

First and foremost, it is important to understand that regular maintenance is necessary to keep your solar system running optimally. While solar panels are designed to be durable and withstand various weather conditions, they are still exposed to the elements, such as rain, wind, and dust. Over time, this can result in the accumulation of debris on the panels, reducing their efficiency. Therefore, it is advisable to clean the panels periodically, especially in areas with high levels of pollution or dust.

In addition to cleaning, it is essential to inspect the solar system regularly for any signs of damage or wear. This includes checking for loose connections, cracks in the panels, or any other visible issues. By identifying and addressing these problems early on, you can prevent further damage and ensure the system continues to generate electricity efficiently.

Another critical aspect of solar system maintenance is monitoring its performance. Most modern solar energy systems come equipped with monitoring software that allows you to track energy production and identify any deviations from normal performance. By keeping an eye on your system's output, you can quickly identify and address any potential issues, such as a faulty inverter or a malfunctioning panel.

It is also worth mentioning the importance of professional inspections and maintenance. While you can perform basic maintenance tasks yourself, it is advisable to have a professional solar technician conduct a thorough inspection at least once a year. They have the expertise and specialized tools to identify any underlying problems that may not be visible to the untrained eye.

Regular maintenance not only ensures the optimal performance of your solar system but also extends its lifespan. By investing time and effort into maintaining your solar energy system, you can continue to reap the benefits of clean and renewable energy for many years to come.

In conclusion, solar system maintenance is an essential aspect of harnessing the power of solar energy. By following proper maintenance procedures, such as cleaning, inspecting, monitoring, and seeking professional assistance when needed, you can ensure the longevity and efficiency of your solar energy system. Regular maintenance not only maximizes the benefits of solar energy but also contributes to a sustainable future for all.

Regular Cleaning and Inspection

Regular cleaning and inspection are crucial aspects of maintaining the efficiency and longevity of solar energy systems in commercial buildings. By regularly performing these tasks, you can ensure that your solar panels are operating at their optimal capacity, maximizing energy generation and reducing the need for costly repairs or replacements.

Cleaning solar panels on a regular basis is essential to remove dirt, dust, and other debris that can accumulate on their surfaces. These particles can block sunlight and reduce the panel's ability to convert solar energy into electricity. Regular cleaning can be done using a soft brush or sponge with mild soap and water. Avoid using abrasive materials or harsh chemicals that could damage the panels. It is also advisable to clean the panels early in the morning or late in the evening to prevent water from evaporating too quickly and leaving behind mineral deposits.

In addition to cleaning, regular inspection of the entire solar energy system is necessary to identify any potential issues or malfunctions. This inspection should include checking for loose or damaged wiring, loose connections, and signs of corrosion. By conducting these inspections periodically, you can catch problems early on and prevent them from escalating into larger, more costly repairs. It is recommended to hire a professional solar energy technician to perform a thorough inspection at least once a year.

By practicing regular cleaning and inspection, you can ensure that your solar energy system remains efficient and reliable. This will contribute to the overall success of your commercial building's energy savings and sustainability goals.

Furthermore, it is essential to keep a record of all cleaning and inspection activities. By maintaining a detailed log, you can track the frequency of cleaning and inspection, note any repairs or maintenance performed, and identify any patterns or recurring issues. This log can serve as a valuable resource for future reference and help you plan and budget for future maintenance needs.

In conclusion, regular cleaning and inspection are vital for the optimal performance and longevity of solar energy systems in commercial buildings. By incorporating these practices into your maintenance routine, you can maximize energy generation, reduce costs, and contribute to a more sustainable future. Remember to consult with professionals and adhere to manufacturer guidelines to ensure safe and effective cleaning and inspection procedures.

Monitoring System Performance

Solar energy systems are a reliable and sustainable solution for commercial buildings, offering numerous benefits such as reduced energy costs, lower carbon emissions, and increased energy independence. However, it is crucial to closely monitor the performance of these systems to maximize their efficiency and ensure optimal operation.

Monitoring system performance involves analyzing various metrics to track the functionality and output of solar energy systems. By regularly monitoring these systems, building owners and operators can identify and address any issues promptly, preventing potential disruptions and maximizing the return on investment.

One key metric to monitor is the energy production of the solar panels. This can be measured by comparing the actual energy generated by the system to the expected energy output based on the system's capacity and environmental factors. By tracking this metric, any underperformance or inefficiency in the system can be identified and rectified.

Additionally, monitoring the performance of inverters is crucial. Inverters convert the direct current (DC) energy generated by the solar panels into alternating current (AC) energy used in commercial buildings. By monitoring inverter performance, any malfunctions or efficiency losses can be detected and addressed promptly, ensuring the system operates optimally.

Furthermore, monitoring battery performance is essential for solar energy systems that incorporate energy storage. Batteries store excess energy generated by the solar panels for use during periods of low or no sunlight. Regularly monitoring battery performance helps ensure their longevity, efficiency, and capacity to provide uninterrupted power supply when needed.

Monitoring system performance also involves analyzing weather conditions and their impact on energy production. By tracking weather patterns and correlating them with energy output, building owners can identify any anomalies or fluctuations in performance caused by external factors and take appropriate actions.

To effectively monitor system performance, various technological solutions are available, ranging from cloud-based monitoring platforms to on-site data loggers. These tools provide real-time data on key performance metrics, allowing building owners and operators to make informed decisions and take proactive measures to optimize system performance.

In conclusion, monitoring system performance is essential in the solar energy industry. By closely tracking metrics such as energy production, inverter performance, battery health, and weather conditions, building owners and operators can ensure their solar energy systems operate efficiently, minimize downtime, and maximize the benefits of solar energy for commercial buildings.

Troubleshooting Common Issues

When it comes to harnessing the power of solar energy for commercial buildings, it is essential to be prepared for any potential issues that may arise. This subchapter aims to provide guidance on troubleshooting common issues encountered in solar energy systems. Whether you are an industry professional, a building owner, or simply someone interested in solar energy solutions, this section will equip you with the knowledge to address and resolve problems effectively.

One of the most frequent issues in solar energy systems is a decline in energy production. Several factors can contribute to this problem, including shading from nearby buildings or trees, soiling of the solar panels, or faulty electrical connections. In this subchapter, we will explore these issues in detail, offering practical solutions to maximize energy output. For instance, we will discuss the importance of regular cleaning and maintenance to ensure optimal performance of solar panels, as well as strategies to mitigate shading effects.

Another common issue that can occur is the malfunction of inverters, which are responsible for converting the direct current (DC) produced by solar panels into alternating current (AC) for use in buildings. In this subchapter, we will guide you through troubleshooting steps to identify and address inverter problems effectively. This will include techniques to diagnose faults, such as checking for error codes and monitoring system performance.

Furthermore, this subchapter will address potential issues related to battery storage systems, which are often integrated into commercial solar energy solutions. We will explore concerns such as battery degradation, improper sizing, and charging/discharging problems. By understanding these issues and their resolutions, you will be better equipped to optimize the performance and longevity of your solar energy system.

Finally, we will touch upon safety considerations when troubleshooting solar energy systems. We will emphasize the importance of following safety protocols, such as shutting off the system before conducting any inspections or repairs. Additionally, we will provide guidelines on when to seek professional assistance to ensure the safety and effectiveness of troubleshooting efforts.

In conclusion, this subchapter serves as a comprehensive guide to troubleshooting common issues encountered in solar energy systems for commercial buildings. By addressing topics such as declining energy production, inverter malfunctions, battery storage problems, and safety considerations, we aim to empower everyone with the knowledge and skills necessary to overcome challenges and maximize the benefits of solar energy solutions.

Chapter 6: Case Studies: Successful Solar Implementations in Commercial Buildings

Office Buildings

Office buildings are a vital part of our modern society, serving as the central hub for businesses, organizations, and institutions. However, these buildings are also one of the largest consumers of energy, contributing to greenhouse gas emissions and straining our natural resources. In recent years, there has been a growing recognition of the need to find sustainable and renewable energy solutions to power these commercial buildings.

Solar energy has emerged as a leading solution for powering office buildings, offering numerous benefits for both the environment and building owners. By harnessing the power of the sun, office buildings can reduce their dependence on traditional energy sources and significantly lower their carbon footprint. Solar panels installed on rooftops or parking structures can generate clean and renewable electricity, providing a reliable source of energy all year round.

The adoption of solar energy in office buildings not only contributes to a greener planet but also offers financial advantages. Installing solar panels can result in substantial cost savings on electricity bills, as well as potential tax incentives and grants offered by government agencies. Additionally, solar energy systems require minimal maintenance, making them a cost-effective long-term investment.

Another advantage of solar energy in office buildings is its potential to enhance the building's overall value. With sustainability becoming a key consideration in real estate, commercial properties equipped with solar energy systems are in high demand. Businesses and tenants are increasingly looking for environmentally responsible spaces, and a solar-powered office building can provide a competitive edge in the market.

Furthermore, solar energy solutions for office buildings can be tailored to meet specific energy needs. Depending on the building's size, location, and energy requirements, solar panels can be installed in various configurations, allowing for maximum efficiency and energy output. Battery storage systems can also be integrated to store excess energy generated during the day for use during peak demand periods or in the event of a power outage.

In conclusion, solar energy solutions have the potential to revolutionize the way office buildings are powered. By harnessing the sun's abundant energy, these buildings can become more sustainable, cost-effective, and environmentally friendly. As the demand for renewable energy continues to grow, embracing solar power in office buildings is a crucial step towards a greener and brighter future for all.

Energy Efficiency Improvements

In today's world, where the demand for energy is increasing rapidly, finding sustainable and efficient solutions is more important than ever. The subchapter "Energy Efficiency Improvements" aims to shed light on the various ways in which solar energy can be harnessed to power commercial buildings, leading to a significant reduction in energy consumption.

Solar energy is a clean and renewable source of power that holds immense potential for addressing our energy needs. By implementing solar solutions in commercial buildings, we can not only reduce our carbon footprint but also save valuable resources and money.

One of the key aspects of energy efficiency improvements is the installation of solar panels. These panels convert sunlight into electricity, which can be used to power lighting, heating, ventilation, and air conditioning systems, among other things. By utilizing solar panels, commercial buildings can significantly decrease their dependence on traditional energy sources, such as fossil fuels. This not only reduces greenhouse gas emissions but also helps to combat climate change.

Another important factor to consider for energy efficiency improvements is the use of energy-efficient appliances and technologies. By replacing outdated and energy-guzzling equipment with energy-efficient alternatives, commercial buildings can further optimize their energy consumption. This can include installing LED lighting, using smart sensors for lighting and HVAC systems, and employing high-efficiency heating and cooling systems. These improvements not only save energy but also result in substantial cost savings for building owners and occupants.

Moreover, energy management systems play a crucial role in monitoring and optimizing energy consumption. These systems provide real-time data on energy usage, allowing building managers to identify areas where energy is being wasted and make necessary adjustments. By implementing effective energy management systems, commercial buildings can track their energy consumption patterns, set energy-saving goals, and even automate energy usage during peak times.

Energy efficiency improvements not only benefit the environment but also enhance the reputation of commercial buildings. By demonstrating a commitment to sustainability and clean energy, businesses can attract environmentally conscious consumers and gain a competitive edge in the market.

In conclusion, the subchapter "Energy Efficiency Improvements" highlights the significant role solar energy plays in powering commercial buildings. By utilizing solar panels, energy-efficient appliances, and effective energy management systems, commercial buildings can reduce their energy consumption, save money, and contribute to a greener future for all.

Financial Savings and ROI

Financial Savings and ROI in Solar Energy Solutions for Commercial Buildings

In today's rapidly evolving world, where the need for sustainable energy solutions is more critical than ever, solar energy has emerged as a game-changer. As the world transitions towards cleaner and greener energy sources, commercial buildings can play a pivotal role in driving this progress. Solar energy solutions not only contribute to a more sustainable future but also offer substantial financial savings and return on investment (ROI) for businesses.

The adoption of solar energy solutions in commercial buildings brings forth numerous financial benefits. One of the most significant advantages is the reduction of energy costs. By harnessing the power of the sun, businesses can generate their own electricity, thus reducing their reliance on traditional energy sources. This independence from the grid leads to significant savings on monthly utility bills, allowing businesses to allocate those funds towards other critical areas of their operations.

Moreover, solar energy solutions offer long-term financial benefits. Commercial buildings equipped with solar panels become self-sufficient powerhouses, generating electricity for decades to come. The initial investment in solar panels can be recouped over time through the savings realized on electricity costs. As energy prices continue to rise, the financial benefits of solar energy solutions become even more compelling, ensuring a positive ROI in the long run.

In addition to cost savings, solar energy solutions also provide businesses with the opportunity to generate revenue. Many governments and utility companies offer incentives and feed-in tariffs for businesses that produce excess energy and feed it back into the grid. This enables businesses to not only reduce their own energy bills but also earn money by selling surplus energy.

Furthermore, solar energy solutions enhance the value of commercial buildings. As sustainability becomes an increasingly important criterion for businesses and customers, buildings equipped with solar panels gain a competitive edge in the market. This added value can translate into higher property prices and increased demand for commercial spaces.

In conclusion, financial savings and ROI are crucial aspects of solar energy solutions for commercial buildings. By reducing energy costs, providing long-term financial benefits, enabling revenue generation, and enhancing property value, solar energy solutions offer a compelling financial case for businesses. Embracing solar energy not only contributes to a more sustainable future but also provides tangible financial advantages, making it a win-win situation for every business and individual.

Environmental Impact Reduction

Solar energy is a powerful solution that can significantly reduce the environmental impact of commercial buildings. In this subchapter, we will explore the various ways in which solar energy can contribute to environmental sustainability and the benefits it brings to everyone.

One of the most significant environmental benefits of solar energy is its ability to reduce greenhouse gas emissions. Unlike traditional energy sources such as fossil fuels, solar energy generates electricity without producing harmful emissions. By switching to solar power, commercial buildings can play a crucial role in mitigating climate change and reducing air pollution. This not only benefits the environment but also improves the overall quality of the air we breathe.

Another aspect of environmental impact reduction through solar energy is its role in conserving natural resources. Solar panels harness the power of the sun, a renewable energy source that will never run out. By utilizing this abundant resource, commercial buildings can decrease their reliance on finite resources like coal, oil, and natural gas. This reduction in resource consumption helps to preserve natural habitats, protect biodiversity, and mitigate the negative impacts of resource extraction.

Solar energy also contributes to water conservation, which is a critical environmental concern in many regions. Traditional power generation methods, such as thermal power plants, require vast amounts of water for cooling purposes. By contrast, solar energy systems require little to no water for operation. By adopting solar power, commercial buildings can significantly reduce their water footprint, easing the strain on local water supplies and supporting sustainable water management practices.

In addition to these direct environmental benefits, solar energy also has indirect positive impacts on the environment. The widespread adoption of solar power drives innovation and technological advancements, leading to more efficient and sustainable energy solutions. Moreover, solar energy systems can be integrated with other green building practices, such as energy-efficient lighting, smart heating and cooling systems, and sustainable materials, further reducing the environmental footprint of commercial buildings.

By embracing solar energy, every individual and commercial building can play a crucial role in mitigating the environmental challenges we currently face. Whether you are a building owner, investor, or simply a concerned citizen, embracing solar energy solutions for commercial buildings is an opportunity to contribute to a cleaner, healthier, and more sustainable future for everyone. Together, we can power progress and make a positive impact on our environment.

Retail Establishments

In recent years, there has been a growing awareness and demand for sustainable practices in various industries, and the retail sector is no exception. As consumers become more conscious of their environmental impact, retail establishments are under increasing pressure to adopt sustainable solutions. One of the most viable and effective options is the integration of solar energy systems.

Solar energy has emerged as a game-changer in the world of renewable energy sources. It harnesses the power of the sun to generate electricity, reducing dependence on fossil fuels and minimizing greenhouse gas emissions. The benefits of solar energy for retail establishments are numerous and far-reaching.

First and foremost, solar energy can significantly reduce operating costs for retail businesses. By generating their own electricity, retail establishments can lower their reliance on the grid, thereby reducing energy bills. Moreover, many governments and utility companies offer financial incentives, such as tax credits and rebates, to encourage the adoption of solar energy. These incentives can further offset the initial investment and lead to long-term savings.

Another advantage of solar energy for retail establishments is its positive brand image. By investing in solar energy, retailers demonstrate their commitment to sustainability and environmental responsibility. This can attract environmentally-conscious consumers who prefer to support businesses that align with their values. Moreover, solar panels on the roof or parking lot can serve as a visible reminder of the retailer's dedication to renewable energy, creating a positive association in customers' minds.

Furthermore, solar energy systems can enhance the resilience of retail establishments. During power outages or emergencies, solar panels equipped with battery storage can continue to provide electricity, ensuring uninterrupted operations. This is particularly important for retailers who rely heavily on refrigeration, lighting, and other essential power-dependent systems.

Lastly, the adoption of solar energy can contribute to a broader shift towards a more sustainable future. By reducing reliance on fossil fuels, retail establishments can play a part in mitigating climate change and reducing air pollution. This not only benefits the environment but also improves the overall health and well-being of communities.

In conclusion, solar energy offers a multitude of benefits for retail establishments. From reducing operating costs to enhancing brand image and resilience, solar power is a smart investment for any retailer looking to create a sustainable and environmentally-friendly business. By embracing solar energy solutions, retail establishments can lead the way in powering progress towards a greener future for all.

Enhanced Brand Image

In an era where sustainability and environmental consciousness are gaining significant importance, businesses across all industries are seeking ways to enhance their brand image. This subchapter will explore the benefits of adopting solar energy solutions for commercial buildings, and how it can contribute to an enhanced brand image for businesses of all types.

Solar energy is a clean, renewable, and abundant source of power that can significantly reduce a business's carbon footprint. By implementing solar energy solutions, companies can demonstrate their commitment to sustainability and environmental responsibility. This sends a powerful message to customers, employees, and stakeholders that the organization is actively contributing to a greener future.

One of the key advantages of solar energy is that it provides businesses with greater energy independence. By generating their own electricity, companies can reduce their reliance on traditional power sources, such as fossil fuels. This not only helps to mitigate the risks associated with fluctuating energy prices but also ensures a more stable and secure energy supply.

Moreover, installing solar panels on commercial buildings can serve as a visible symbol of a company's commitment to renewable energy. The sight of solar panels on rooftops or in parking lots can attract attention and generate positive public perception. This can lead to increased brand recognition, customer loyalty, and even attract new customers who value sustainability.

Additionally, solar energy solutions can provide significant cost savings over the long term. While the initial investment in solar panels and equipment may seem high, businesses can benefit from reduced or even eliminated electricity bills. The money saved can be reinvested in other areas of the business, such as research and development, employee training, or marketing efforts, further enhancing the brand image.

Furthermore, government incentives and tax breaks are often available to businesses that adopt solar energy solutions. These incentives can help offset the initial costs and make solar energy more financially viable. Taking advantage of these incentives not only demonstrates fiscal responsibility but also positions the company as a forward-thinking and socially responsible organization.

In summary, embracing solar energy solutions can have a transformative impact on a business's brand image. By showcasing a commitment to sustainability, energy independence, and cost savings, companies can attract and retain customers who align with their values. Solar energy can be a powerful tool to differentiate a business in a crowded market and position it as a leader in the transition towards a cleaner and more sustainable future.

Customer Engagement

In the rapidly evolving world of solar energy, successful customer engagement is crucial for the growth and adoption of solar solutions in commercial buildings. By understanding and meeting the needs of customers, solar energy providers can build lasting relationships and drive progress towards a sustainable future. This subchapter explores the importance of customer engagement in the context of solar energy solutions for commercial buildings.

Customer engagement goes beyond simply selling solar panels or systems; it involves understanding the unique challenges and requirements of each customer. Every commercial building is different, and solar energy providers must tailor their offerings to meet the specific needs of their customers. This requires active listening, effective communication, and a deep understanding of the benefits and limitations of solar energy.

One key aspect of customer engagement is education. Many commercial building owners and managers may have limited knowledge of solar energy and its potential benefits. By providing clear and concise information, solar energy providers can help customers make informed decisions and overcome any misconceptions or doubts they may have. This can include explaining the financial incentives, environmental benefits, and long-term cost savings associated with solar energy solutions.

Furthermore, customer engagement involves ongoing support and assistance. From the initial consultation to system installation and beyond, solar energy providers should be readily available to address any questions or concerns that may arise. This level of personalized attention builds trust and confidence in the customer-provider relationship, enhancing customer satisfaction and loyalty.

In addition to individual customer engagement, community engagement is also vital. By actively participating in local events, workshops, and educational initiatives, solar energy providers can raise awareness and promote the benefits of solar energy within the community. This fosters a sense of shared responsibility and encourages more commercial buildings to consider solar energy solutions.

Overall, customer engagement is key to driving the adoption of solar energy solutions in commercial buildings. By understanding customer needs, providing education and ongoing support, and actively engaging with the community, solar energy providers can build strong relationships and pave the way for a sustainable future powered by solar energy.

Increased Foot Traffic

One of the key benefits of incorporating solar energy solutions in commercial buildings is the potential to increase foot traffic. As more and more businesses strive to become environmentally conscious, consumers are actively seeking out establishments that align with their values. By adopting solar energy, commercial buildings can attract a wider customer base, boost their reputation, and ultimately drive more sales.

Solar energy is a sustainable and clean energy source that resonates with a diverse audience. In today's world, where climate change and environmental preservation are hot topics, people are becoming increasingly aware of their carbon footprint. They are actively seeking out businesses that prioritize renewable energy sources. By embracing solar energy, commercial buildings can position themselves as leaders in sustainability and attract customers who are eager to support eco-friendly initiatives.

Moreover, the presence of solar panels on a commercial building can act as a powerful marketing tool. The visible display of solar panels sends a strong message to passersby, showcasing the business's commitment to a greener future. This visual cue can pique the interest of potential customers, encouraging them to step inside and learn more about what the establishment has to offer.

Increased foot traffic not only benefits businesses financially but also helps to create a sense of community. When more people visit a commercial building, it fosters interaction and engagement. This can lead to a vibrant atmosphere, creating a positive experience for both customers and employees. Additionally, as foot traffic increases, businesses can explore partnerships with local organizations or host community events, further solidifying their position as a socially responsible entity.

Furthermore, by attracting more customers, commercial buildings can generate additional revenue streams. Solar energy can help reduce operational costs, freeing up funds for marketing initiatives, product development, or expanding services. This, in turn, can help businesses stay competitive and continue to grow in an ever-evolving market.

In conclusion, the integration of solar energy solutions in commercial buildings offers numerous benefits, including increased foot traffic. By embracing renewable energy, businesses can attract a wider customer base, enhance their reputation, and create a sense of community. The visibility of solar panels also acts as a powerful marketing tool, enticing potential customers to explore the offerings of the establishment. Moreover, increased foot traffic can lead to additional revenue streams and business growth. As the demand for environmentally conscious businesses continues to rise, adopting solar energy solutions becomes a strategic move for commercial buildings, appealing to a diverse audience and driving progress towards a sustainable future.

Chapter 7: Overcoming Challenges and Limitations in Solar Energy for Commercial Buildings

Regulatory and Permitting Hurdles

When it comes to adopting solar energy solutions for commercial buildings, one cannot overlook the regulatory and permitting hurdles that often arise. While solar power offers numerous benefits such as cost savings, environmental sustainability, and energy independence, navigating the complex web of regulations and obtaining the necessary permits can pose challenges for businesses and individuals alike.

The regulatory landscape surrounding solar energy varies from region to region, and it is important for every stakeholder in the solar energy niche to understand the specific requirements and restrictions within their jurisdiction. This subchapter aims to shed light on the most common regulatory and permitting hurdles and provide guidance on how to navigate them effectively.

One of the most significant regulatory hurdles facing the solar energy industry is the interconnection process. Connecting solar systems to the grid requires adherence to strict technical standards and safety regulations. Understanding the interconnection process and ensuring compliance with local utility requirements can be a time-consuming and intricate task. However, it is crucial to ensure that solar installations are seamlessly integrated into the existing electric grid to avoid safety hazards and disruption to the power supply.

Another regulatory hurdle is the complex web of building codes and zoning regulations. These codes and regulations dictate the design, construction, and location of solar installations. Understanding and complying with these codes is essential to ensure the safety, reliability, and aesthetic integration of solar systems within the built environment. Moreover, zoning restrictions may limit the size or placement of solar panels, requiring careful planning and coordination with local authorities.

Permitting is yet another challenge that individuals and businesses face when adopting solar energy solutions. Obtaining the necessary permits can be a time-consuming and costly process, involving extensive paperwork, inspections, and fees. It is essential to understand the specific permitting requirements within a particular jurisdiction and to engage with the relevant authorities early in the project planning phase to avoid delays and additional costs.

Despite the regulatory and permitting hurdles, the solar energy industry has made significant strides in streamlining processes and reducing barriers to entry. Many jurisdictions have implemented standardized permitting procedures and online platforms to facilitate the application process. Additionally, solar advocacy groups and industry associations have been instrumental in advocating for simplified regulations and permitting requirements.

In conclusion, regulatory and permitting hurdles can pose significant challenges to the adoption of solar energy solutions for commercial buildings. However, with careful planning, knowledge of local regulations, and engagement with relevant authorities, these hurdles can be effectively overcome. The solar energy industry continues to evolve and innovate, and with the collective efforts of stakeholders, we can work towards a future where solar power is accessible to all, delivering clean and sustainable energy to power progress.

Understanding Local Codes and Regulations

When it comes to implementing solar energy solutions for commercial buildings, understanding local codes and regulations is of utmost importance. These codes and regulations vary from one location to another, and they play a crucial role in determining the feasibility and success of any solar energy project. This subchapter aims to provide a comprehensive overview of the key aspects related to local codes and regulations in the context of solar energy, ensuring that everyone, from novices to experts, can navigate this complex landscape.

Local codes and regulations pertaining to solar energy cover a wide range of areas, including building permits, zoning restrictions, electrical codes, fire safety, and interconnection standards. Building permits are essential for any solar energy installation, as they ensure that the project complies with local building codes, structural requirements, and safety standards. Zoning restrictions, on the other hand, dictate where solar panels can be installed, considering factors such as setbacks, aesthetics, and neighborhood covenants.

Electrical codes are crucial for ensuring the safe and efficient installation of solar energy systems. These codes govern the design, installation, and maintenance of electrical systems, including the connection of solar panels to the electrical grid. Compliance with these codes guarantees the protection of both the building occupants and the utility grid.

Fire safety regulations are designed to minimize the risk of fires caused by solar energy systems. They provide guidelines for proper installation, use of fire-resistant materials, and access for firefighters in case of emergencies. Understanding and adhering to these regulations is vital to mitigate potential risks associated with solar energy installations.

Interconnection standards dictate the rules and procedures for connecting solar energy systems to the electrical grid. These standards ensure that the solar system can safely and seamlessly integrate with the grid, allowing for the export of excess electricity and the import of power when needed. Compliance with interconnection standards facilitates the smooth functioning of solar energy systems and promotes the growth of renewable energy generation.

Additionally, it is important to note that local codes and regulations are constantly evolving to keep pace with the advancements in solar energy technology and the changing needs of communities. Staying updated with these changes is crucial to ensure compliance and maximize the benefits of solar energy solutions.

In conclusion, understanding local codes and regulations is essential for anyone involved in the solar energy industry. Whether you are a business owner, a solar installer, or a policymaker, having a comprehensive understanding of these codes and regulations is vital to ensure the successful implementation of solar energy solutions for commercial buildings. By adhering to these standards, we can accelerate the adoption of solar energy, reduce our dependence on fossil fuels, and create a sustainable future for all.

Navigating Permitting Processes

The journey towards harnessing solar energy for commercial buildings involves more than just installing solar panels on rooftops. It requires going through various permitting processes to ensure compliance with local regulations and building codes. Navigating these processes can be complex and time-consuming, but with the right knowledge and guidance, it can be a smooth and successful experience.

Understanding the permitting processes is crucial for anyone interested in solar energy solutions for commercial buildings. This subchapter will provide an overview of the key steps involved in obtaining the necessary permits, empowering every individual to embark on their solar energy journey.

The first step in navigating the permitting processes is to educate yourself about the local regulations and building codes specific to your area. Familiarizing yourself with these requirements will help you understand the scope of work involved and the permits you need to obtain. It is important to note that these regulations may vary from one jurisdiction to another, so thorough research is essential.

Once you have a clear understanding of the local regulations, the next step is to prepare the necessary documentation for permit applications. This may include architectural drawings, structural calculations, electrical plans, and equipment specifications. Working with a qualified solar energy professional can greatly assist in preparing these documents accurately and efficiently.

After the documentation is complete, it is time to submit your permit applications to the relevant authorities. This step often involves paying fees and waiting for the application to be reviewed. The review process may include inspections, ensuring that the proposed solar energy system meets all safety and structural requirements.

While navigating the permitting processes, it is important to maintain open lines of communication with the permitting authorities. Regular follow-ups can help expedite the review process and address any concerns or questions they may have. Building positive relationships with the permitting officials can make the entire process smoother and more collaborative.

Once the permits are approved, construction can begin. It is crucial to adhere to the approved plans and specifications during the installation process. Regular inspections may occur during construction to ensure compliance with the permits and regulations.

In conclusion, navigating the permitting processes is an essential part of implementing solar energy solutions for commercial buildings. By understanding the local regulations, preparing accurate documentation, and maintaining open communication with permitting authorities, anyone can successfully navigate this journey. With each successful permit obtained, we move one step closer to a future powered by clean and sustainable solar energy.

Dealing with Utility Interconnection Issues

In the world of solar energy, one of the most critical aspects to consider when implementing solar solutions for commercial buildings is utility interconnection. This subchapter aims to shed light on the various challenges and potential solutions associated with this aspect of solar energy integration.

Utility interconnection refers to the process of connecting a solar energy system to the existing electrical grid. While solar energy offers numerous benefits, such as reducing carbon emissions and lowering energy costs, navigating the interconnection process can be complex and time-consuming. However, by understanding the common issues and taking proactive steps, businesses can overcome these obstacles and embrace the power of solar energy.

One of the primary challenges of utility interconnection is dealing with the bureaucratic red tape often associated with utility companies. These organizations have specific rules and regulations that must be followed, which can vary from region to region. This can lead to delays and increased costs if not properly managed. To tackle this issue, it is essential to conduct thorough research and engage with experienced professionals who can guide you through the interconnection process smoothly.

Another common obstacle is the technical compatibility between the solar energy system and the electrical grid. Ensuring that the system is properly designed and integrated with the existing infrastructure is crucial for a successful interconnection. It may require modifications to the electrical system or the installation of additional equipment. Collaboration with qualified solar energy experts is vital to ensure a seamless connection and avoid any potential issues down the line.

Moreover, grid capacity limitations can pose a significant challenge. In some areas, the electrical grid may not have the capacity to accommodate additional solar energy systems. This can result in rejection or lengthy waiting periods for interconnection. However, innovative solutions such as energy storage systems and smart grid technologies can help address these limitations and enable a smoother integration of solar energy.

To overcome these utility interconnection issues, businesses should establish strong communication channels with utility companies, seek professional advice, and conduct thorough site assessments. By understanding the utility interconnection process and addressing potential challenges proactively, commercial buildings can successfully harness the power of solar energy, reduce their carbon footprint, and lower energy costs.

In conclusion, utility interconnection is a critical aspect of solar energy integration for commercial buildings. By being aware of the challenges and taking necessary steps, businesses can overcome the bureaucratic hurdles, address technical compatibility issues, and tackle grid capacity limitations. This subchapter aims to equip the audience with the knowledge and strategies needed to navigate utility interconnection successfully, empowering them to embrace solar energy solutions and contribute to a sustainable future.

Technical Limitations

In the pursuit of harnessing solar energy for commercial buildings, it is important to acknowledge and understand the technical limitations that exist in this field. While solar energy solutions have made remarkable progress in recent years, there are still challenges to be addressed and overcome. This subchapter aims to shed light on these limitations and provide a comprehensive overview for anyone interested in solar energy solutions.

One of the primary technical limitations of solar energy is its intermittent nature. Solar panels can only generate electricity when the sun is shining, which means they are unable to produce power during nighttime or in cloudy weather conditions. This intermittency poses a challenge for commercial buildings that require a consistent and reliable energy source. However, advancements in energy storage technologies, such as batteries, are mitigating this issue by allowing excess energy to be stored for use during non-solar hours.

Another limitation is the efficiency of solar panels. Currently, solar panels have an average efficiency of around 20%, meaning they can convert only a fraction of the sunlight they receive into usable electricity. This efficiency can be affected by various factors, including temperature, shading, and dust accumulation on the panels. Ongoing research and development efforts are focused on improving the efficiency and durability of solar panels to enhance their performance and reduce costs.

The physical space required for solar installations is also a consideration. Commercial buildings often have limited roof space, making it a challenge to accommodate a sufficient number of solar panels. Additionally, the orientation and angle of the roof can impact the effectiveness of solar panels. However, innovative solutions such as solar canopies or ground-mounted systems are being explored to overcome these space limitations.

Integration with existing electrical infrastructure is another technical limitation to be mindful of. Commercial buildings typically have complex electrical systems, and integrating solar energy solutions seamlessly can be a complex task. Ensuring compatibility, stability, and safety requires careful planning and coordination with electrical experts.

Despite these limitations, solar energy solutions for commercial buildings continue to evolve and improve. Technological advancements, coupled with supportive government policies and incentives, are driving the adoption of solar energy on a larger scale. As research and development efforts progress, we can expect to see more efficient, affordable, and seamlessly integrated solar energy solutions for commercial buildings.

In conclusion, while technical limitations exist in the field of solar energy, ongoing advancements and innovation are steadily overcoming these challenges. By understanding these limitations and actively working towards their resolution, we can power progress and accelerate the adoption of solar energy solutions for commercial buildings.

Space Constraints and Panel Orientation

When it comes to harnessing the power of solar energy for commercial buildings, space constraints and panel orientation play a crucial role. In this subchapter, we will explore how these factors can impact the efficiency and effectiveness of solar energy systems in commercial settings.

Space constraints refer to the limited availability of roof or ground space for installing solar panels. Commercial buildings often have various structures, such as HVAC systems, vents, and skylights, which can limit the available space for solar panel installation. However, with advancements in solar technology, there are now innovative solutions that can help overcome these constraints.

One such solution is the use of building-integrated photovoltaics (BIPV). BIPV systems integrate solar panels directly into the building's architecture, such as windows, walls, or roof tiles. By utilizing these surfaces, commercial buildings can maximize their solar energy generation without compromising available space. BIPV not only generates electricity but also enhances the building's aesthetics and reduces the need for traditional building materials.

Another consideration for solar energy systems is panel orientation. The orientation of solar panels determines the amount of sunlight they receive throughout the day, directly impacting their energy production. In the northern hemisphere, it is generally recommended to orient solar panels towards the south to maximize sunlight exposure. However, this may not always be feasible due to building orientation or shading from surrounding structures.

To overcome these challenges, solar energy systems can incorporate tracking technology. Solar trackers follow the sun's path throughout the day, adjusting the panels' angle to ensure optimal sunlight absorption. This technology can significantly enhance the energy production of solar panels, making them a viable option even in less ideal orientations.

Moreover, it is essential to consider the tilt angle of solar panels. The tilt angle determines how well the panels capture sunlight throughout the year, accounting for the sun's varying position. By adjusting the tilt angle based on the geographical location, commercial buildings can optimize their solar energy production and maximize their return on investment.

In conclusion, space constraints and panel orientation are critical factors to consider when implementing solar energy solutions for commercial buildings. With innovative approaches like BIPV and solar tracking technology, it is possible to overcome space limitations and maximize energy production. By understanding the importance of panel orientation and tilt angle, commercial buildings can effectively harness the power of solar energy and contribute to a sustainable future for everyone.

Intermittent Power Generation

In the world of solar energy, intermittent power generation refers to the fluctuation in energy output from solar panels due to various factors. While solar energy is undoubtedly one of the most abundant and sustainable sources of power, it is important to understand the intermittent nature of this energy generation method.

Solar panels rely on sunlight to produce electricity. However, the availability of sunlight is influenced by factors such as weather conditions, time of day, and geographical location. As a result, solar power generation can experience periods of high output and low output, leading to intermittency.

One of the main challenges with intermittent power generation is the inability to consistently meet the energy demands of commercial buildings. This can be particularly significant during peak usage hours when businesses require a steady and reliable energy supply. To address this issue, advanced energy storage systems are often employed to store excess energy generated during high output periods. These stored reserves can then be utilized during low output periods, ensuring a more continuous and reliable power supply.

Additionally, grid interconnection plays a crucial role in mitigating the impact of intermittent power generation. Solar energy systems can be connected to the electrical grid, allowing excess energy to be fed back into the grid during periods of high output. This not only ensures that the energy generated is not wasted but also provides an opportunity for commercial buildings to earn credits or financial incentives through net metering programs.

Furthermore, advancements in technology have led to the development of innovative solutions to enhance solar power generation. For instance, the integration of smart grid technologies and predictive analytics can help optimize power generation and consumption, minimizing the impact of intermittency. By analyzing historical weather patterns and energy usage data, these systems can predict when low output periods are likely to occur and adjust energy consumption accordingly.

Intermittent power generation is an inherent characteristic of solar energy systems. However, with the right strategies and technologies in place, its impact on commercial buildings can be effectively managed. By utilizing energy storage, grid interconnection, and advanced analytics, businesses can ensure a more consistent and reliable solar power supply, ultimately driving progress towards a sustainable future.

Battery Storage Solutions

Battery storage solutions have emerged as a game-changer in the field of solar energy, revolutionizing the way commercial buildings harness and utilize renewable power. As the demand for clean energy continues to rise, businesses are increasingly turning to solar power as a sustainable and cost-effective alternative. However, one of the biggest challenges associated with solar energy is its intermittent nature. This is where battery storage solutions come into play.

Battery storage systems allow commercial buildings to store excess energy generated by solar panels during peak production times and use it later when sunlight is not available. By capturing and storing surplus energy, businesses can ensure a steady and reliable power supply, even during periods of low solar generation or grid outages. This not only enhances the overall reliability of the building's power system but also reduces dependence on the traditional electrical grid.

The benefits of battery storage solutions extend beyond providing a consistent power supply. They also offer significant financial advantages. By storing excess solar energy and using it during periods of high electricity demand or peak utility rates, businesses can effectively reduce their energy bills. Additionally, battery storage systems can participate in demand response programs, where energy providers incentivize businesses to reduce their electricity consumption during peak load times. This further contributes to cost savings and allows businesses to play an active role in grid stabilization and efficient energy management.

Furthermore, battery storage solutions enhance the overall flexibility and resilience of commercial buildings. They can be integrated with other renewable energy sources, such as wind or hydro power, to create a comprehensive and diversified energy portfolio. This enables businesses to optimize their energy usage and reduce their carbon footprint, contributing to a greener and more sustainable future.

In recent years, advancements in battery technology have made storage solutions more efficient, affordable, and durable. Lithium-ion batteries, in particular, have become the preferred choice for commercial buildings due to their high energy density, long lifespan, and fast charging capabilities. Moreover, the decreasing costs of batteries and government incentives have made battery storage solutions increasingly accessible to businesses of all sizes.

As the world transitions towards a clean energy future, battery storage solutions are poised to play a pivotal role in enabling the widespread adoption of solar energy in commercial buildings. Embracing these innovative solutions not only benefits businesses financially but also contributes to a more reliable, resilient, and sustainable energy ecosystem.

Chapter 8: The Future of Solar Energy for Commercial Buildings

Technological Advancements

In recent years, the solar energy industry has witnessed significant technological advancements that have revolutionized the way we harness and utilize this abundant renewable resource. These advancements have not only made solar energy more efficient and cost-effective but have also opened up new possibilities for its integration into commercial buildings. In this subchapter, we will explore some of the most notable technological advancements in solar energy and their impact on powering progress in commercial buildings.

One of the key advancements in solar energy technology is the development of high-efficiency solar panels. Traditional solar panels were limited in their ability to convert sunlight into electricity, but with the advent of new materials and designs, modern panels can now achieve conversion efficiencies of over 20%. These high-efficiency panels not only generate more electricity from the same amount of sunlight but also take up less space, making them ideal for installation on commercial rooftops.

Another significant advancement is the integration of solar energy storage solutions. Traditionally, solar energy systems were only able to generate electricity during daylight hours. However, with the introduction of advanced battery technologies, excess solar energy can now be stored for use during periods of low sunlight or high demand. This allows commercial buildings to rely on solar power even when the sun is not shining, further reducing their dependence on grid electricity and lowering operational costs.

Furthermore, advancements in solar tracking systems have greatly improved the overall performance of solar energy systems. Solar trackers are devices that move solar panels to follow the sun's path throughout the day, maximizing their exposure to sunlight. By constantly adjusting the panel's position, solar trackers can increase energy production by up to 25%, making them a valuable addition to commercial solar installations.

Moreover, the development of smart grid technologies has enabled better integration of solar energy into commercial buildings. Smart grids allow for two-way communication between the utility company and the building's solar energy system, enabling real-time monitoring and control of electricity generation and consumption. This not only optimizes energy usage but also facilitates the seamless integration of solar power into the existing electrical infrastructure.

In conclusion, technological advancements in solar energy have significantly transformed the way we harness and utilize this renewable resource. From high-efficiency solar panels and energy storage solutions to solar tracking systems and smart grid technologies, these advancements have propelled the adoption of solar energy in commercial buildings. As we continue to innovate and improve upon existing technologies, the future of solar energy looks brighter than ever, offering sustainable and cost-effective solutions for powering progress in commercial buildings.

Increased Efficiency in Solar Panels

Solar energy has emerged as a leading source of renewable energy, offering numerous benefits for both the environment and businesses. One of the key factors contributing to the widespread adoption of solar energy is the increased efficiency in solar panels. This subchapter explores the advancements in solar panel technology that have led to improved efficiency, making solar energy an even more attractive option for commercial buildings.

Over the years, researchers and engineers have been tirelessly working to enhance the efficiency of solar panels. This has resulted in significant breakthroughs, allowing solar panels to convert sunlight into electricity at a much higher rate than ever before. The efficiency of solar panels is measured by the percentage of sunlight that is converted into usable electricity. Earlier generations of solar panels had an efficiency rate of around 15-20%, but recent innovations have pushed this figure to over 25%.

One of the key advancements in solar panel technology is the use of multi-junction or tandem cells. These cells incorporate multiple layers of semiconductors, each designed to capture a specific portion of the solar spectrum. By optimizing the absorption of different wavelengths of light, multi-junction cells significantly boost the efficiency of solar panels. This technology has primarily been used in space applications but is now being adapted for commercial buildings.

Another promising innovation is the use of perovskite materials in solar panels. Perovskite-based solar cells have shown great potential due to their low cost, flexibility, and high efficiency. Although still in the research and development phase, perovskite solar cells have achieved remarkable efficiency rates, surpassing traditional silicon-based solar cells. This breakthrough could revolutionize the solar energy industry, making solar panels more affordable and accessible for commercial buildings.

Furthermore, advancements in panel design and manufacturing processes have also contributed to increased efficiency. Thin-film solar panels, for example, are lighter, more flexible, and less expensive to produce compared to traditional silicon-based panels. These panels can be easily integrated into building materials, such as windows or roofing, maximizing the use of available space and improving overall efficiency.

The increased efficiency of solar panels has significant implications for commercial buildings. Higher efficiency means that solar panels can generate more electricity from the same amount of sunlight, reducing the reliance on the grid and lowering energy costs. Additionally, the improved performance of solar panels allows businesses to install smaller systems while still meeting their energy needs.

In conclusion, the advancements in solar panel technology have led to increased efficiency, making solar energy an appealing option for commercial buildings. The use of multi-junction cells, perovskite materials, and innovative panel designs have significantly improved the conversion of sunlight into electricity. These advancements not only reduce costs but also pave the way for a more sustainable future. By harnessing the power of the sun more effectively, businesses can contribute to reducing carbon emissions and creating a cleaner environment for everyone.

Integration with Smart Building Systems

In today's rapidly evolving world, the integration of solar energy solutions with smart building systems has emerged as a groundbreaking approach to powering commercial buildings. This innovative synergy not only optimizes energy consumption but also ensures a sustainable future for our planet. In this subchapter, we will explore the numerous benefits and possibilities that arise from the integration of solar energy with smart building systems, catering to the diverse interests of solar energy enthusiasts and professionals alike.

The integration of solar energy systems with smart building technologies enables seamless communication and collaboration between various components of a building's infrastructure. By harnessing the power of solar energy, buildings can become self-sufficient, reducing their reliance on traditional energy sources and minimizing their carbon footprint. Smart building systems, on the other hand, leverage advanced automation and data analytics to enhance energy management, improve operational efficiency, and create a comfortable environment for occupants.

One of the key advantages of integrating solar energy with smart building systems is the ability to monitor and control energy consumption in real-time. Smart meters and sensors provide valuable insights into energy usage patterns, allowing building owners and occupants to make informed decisions about energy conservation. This integration also enables load management, where excess energy generated by solar panels can be stored in batteries or redirected to power other aspects of the building, such as heating, ventilation, and air conditioning (HVAC) systems.

Furthermore, the integration of solar energy with smart building systems promotes predictive maintenance and fault detection. By continuously monitoring the performance of solar panels and other energy-related equipment, potential issues can be identified and resolved proactively, minimizing downtime and reducing maintenance costs. Additionally, the data collected by smart building systems can be analyzed to optimize energy efficiency, identify trends, and implement targeted energy-saving strategies.

For solar energy enthusiasts, the integration with smart building systems opens up a world of possibilities. It allows for the seamless integration of solar panels into building facades or windows, expanding the potential for solar energy generation. Moreover, advanced technologies such as Internet of Things (IoT) and artificial intelligence (AI) can be leveraged to create intelligent energy management systems that not only optimize energy usage but also adapt to changing environmental conditions.

In conclusion, the integration of solar energy solutions with smart building systems represents a significant milestone in the quest for sustainable and efficient energy consumption. It caters to the interests of solar energy enthusiasts and professionals, offering a myriad of benefits such as energy conservation, real-time monitoring, predictive maintenance, and enhanced efficiency. By embracing this integration, we can power progress and pave the way for a brighter, greener future for all.

Emerging Solar Technologies

In recent years, the field of solar energy has witnessed significant advancements and breakthroughs. These emerging solar technologies hold great promise in revolutionizing the way we harness and utilize the power of the sun. In this subchapter, we will explore some of the most exciting developments in the world of solar energy.

One of the most prominent emerging technologies is solar photovoltaic (PV) cells. These cells are designed to convert sunlight directly into electricity. While traditional silicon-based PV cells have been widely used for decades, new materials such as perovskite and thin-film solar cells are gaining attention due to their higher efficiency and lower manufacturing costs. The potential of these emerging PV technologies is immense, as they could make solar power more accessible and affordable for a wider range of applications.

Another exciting avenue in solar energy is concentrated solar power (CSP). Unlike PV cells, CSP systems use mirrors or lenses to concentrate sunlight onto a receiver, which then converts the solar energy into heat or electricity. This technology allows for the storage of solar energy, enabling the generation of electricity even when the sun is not shining. With advancements in CSP, we can expect more efficient systems and increased integration with energy storage solutions, making solar power a more reliable and consistent source of energy.

In addition to PV and CSP, emerging solar technologies encompass innovative concepts such as solar windows, solar paint, and solar textiles. Solar windows, for instance, incorporate transparent solar cells into the glass, allowing buildings to generate electricity without sacrificing natural light. Solar paint, on the other hand, utilizes specialized dyes and nanoparticles to capture solar energy, transforming any surface into a potential energy generator. Solar textiles integrate solar cells into fabrics, opening up possibilities for wearable solar devices and power-generating clothing.

The emergence of these new solar technologies is driving the growth and adoption of solar energy in commercial buildings. With increased efficiency, lower costs, and improved aesthetics, solar power is becoming an attractive option for businesses seeking to reduce their carbon footprint and energy costs. Moreover, these developments have the potential to transform the way we think about energy generation, paving the way for a brighter and more sustainable future.

Whether you are an individual interested in solar energy, a business owner considering solar installations, or a researcher exploring the frontiers of renewable energy, understanding emerging solar technologies is essential. By staying informed about the latest advancements, we can actively contribute to the transition towards a cleaner and more sustainable energy landscape.

Government Policies and Initiatives

In recent years, there has been a growing recognition of the urgent need to transition from fossil fuels to more sustainable and renewable energy sources. Solar energy has emerged as a promising solution, offering numerous benefits such as reduced carbon emissions, cost savings, and energy independence. In this subchapter, we will delve into the various government policies and initiatives that have been put in place to support and promote the adoption of solar energy solutions in commercial buildings.

Governments around the world have realized the potential of solar energy and have introduced a range of policies and initiatives to incentivize its usage. One such policy is the implementation of feed-in tariffs (FITs), which provide financial incentives to businesses and individuals who generate solar energy and feed it back into the grid. These tariffs ensure a fixed, long-term payment for the electricity produced, making solar energy a lucrative investment for commercial building owners.

Another crucial initiative is the provision of subsidies and grants for the installation of solar panels. Governments understand that the upfront costs of solar energy systems can be a significant barrier for many businesses. By offering financial support, they aim to make solar energy more accessible and affordable, encouraging wider adoption.

Furthermore, governments have introduced net metering policies that allow commercial buildings to offset their electricity bills by exporting excess solar energy to the grid. Net metering enables businesses to not only reduce their dependence on the grid but also earn credits for the surplus energy they generate. This feature has proven to be a major driver for solar energy adoption, as it allows businesses to see a tangible return on their investment.

In addition to financial incentives, governments are also taking regulatory measures to facilitate the integration of solar energy into commercial buildings. Building codes and regulations are being updated to include provisions for solar panel installations, making it easier for businesses to navigate the process and ensuring the safety and efficiency of the systems.

Overall, the government policies and initiatives discussed in this subchapter are instrumental in accelerating the adoption of solar energy solutions in commercial buildings. By providing financial incentives, regulatory support, and technical guidance, governments are paving the way for a sustainable future powered by solar energy. Whether you are a business owner, an investor, or simply an individual concerned about the environment, understanding these policies and initiatives is crucial as we collectively work towards a cleaner, greener future for all.

Support for Renewable Energy

Renewable energy sources, such as solar power, offer tremendous potential for addressing the energy crisis and mitigating the harmful effects of climate change. In this subchapter, we will delve into the various forms of support available for renewable energy, with a particular focus on solar energy solutions for commercial buildings. This content aims to bring awareness and understanding to a broad audience, from individuals interested in sustainable living to professionals in the solar energy industry.

Governments worldwide have recognized the importance of transitioning to renewable energy and have implemented numerous support mechanisms. These include financial incentives, policy frameworks, and research and development initiatives. Financial incentives, such as tax credits, grants, and rebates, encourage individuals and businesses to invest in renewable energy systems like solar panels. These incentives help to offset the initial costs of installation and make solar energy more accessible to a wider range of consumers.

Policy frameworks play a crucial role in supporting renewable energy. Governments can set renewable energy targets, implement feed-in tariffs, and establish net metering policies. These measures incentivize the production and consumption of renewable energy, ensuring a stable market for solar power. Additionally, governments can enact regulations and building codes that require or encourage the integration of solar energy systems in commercial buildings, further driving the adoption of sustainable practices.

Research and development initiatives are vital for advancing solar energy technologies. Governments and organizations invest in research to improve solar panel efficiency, energy storage solutions, and grid integration. These advancements not only enhance the overall performance of solar energy systems but also contribute to reducing costs, making solar power more affordable and competitive with traditional energy sources.

Furthermore, the support for renewable energy extends beyond government initiatives. Community-based organizations, environmental groups, and businesses committed to sustainability often provide educational programs, workshops, and resources to promote solar energy adoption. These initiatives empower individuals and businesses to make informed decisions about utilizing solar power and inspire community-wide engagement in renewable energy solutions.

In conclusion, the support for renewable energy, particularly solar power, is essential for driving the transition to a sustainable and clean energy future. Governments, organizations, and communities play significant roles in providing financial incentives, establishing policy frameworks, investing in research and development, and promoting awareness and education. As individuals and professionals, we must seize the opportunities and support renewable energy solutions to power progress towards a greener world.

Incentives for Commercial Solar Adoption

One of the main challenges faced by commercial buildings in adopting solar energy solutions is the initial cost involved in installing solar panels. However, governments, organizations, and utilities around the world have recognized the importance of transitioning to renewable energy sources and have introduced various incentives to encourage commercial solar adoption. These incentives not only help businesses reduce their carbon footprint but also offer significant financial benefits.

One of the most popular incentives is the Investment Tax Credit (ITC). Under the ITC, businesses can claim a tax credit equal to a percentage of the total cost of their solar installation. In the United States, for example, businesses can receive a tax credit of 26% of the total project cost, significantly reducing the upfront investment required. This incentive has been instrumental in promoting commercial solar adoption and has led to a rapid increase in solar installations across the country.

Another incentive that has gained traction is net metering. Net metering allows businesses to offset their electricity bills by exporting excess solar energy back to the grid. This means that during periods of low energy consumption, such as weekends or holidays, businesses can accrue credits that can be used to offset future electricity consumption. Net metering not only reduces electricity costs but also provides businesses with a reliable source of income by selling surplus energy back to the grid.

Many governments and utilities also offer grants and rebates to businesses that install solar panels. These financial incentives can cover a significant portion of the installation cost, making solar energy a viable and attractive option for commercial buildings. Additionally, some states and municipalities offer property tax exemptions or reductions for commercial buildings that invest in solar energy systems. These incentives further contribute to the financial viability of solar adoption for businesses.

In conclusion, incentives play a crucial role in promoting commercial solar adoption. By reducing upfront costs, providing tax credits, offering net metering programs, and providing grants and rebates, governments and utilities are encouraging businesses to transition to renewable energy sources. These incentives not only help businesses reduce their environmental impact but also provide substantial financial benefits. With the increasing availability of incentives, commercial buildings are now better positioned to embrace solar energy and contribute to a more sustainable future.

Potential Impact on the Market

The rapid growth and adoption of solar energy solutions in commercial buildings have the potential to revolutionize the market. The impact of solar energy on the market is far-reaching, affecting various stakeholders and industries. In this subchapter, we will explore the potential impacts of solar energy on the market and how it can benefit everyone.

One of the most significant impacts of solar energy on the market is its potential to reduce dependence on traditional fossil fuels. As the world faces the challenges of climate change and dwindling fossil fuel reserves, solar energy provides a clean and renewable alternative. By harnessing the power of the sun, commercial buildings can significantly reduce their carbon footprint and contribute to a greener future.

The adoption of solar energy solutions in commercial buildings can also lead to cost savings. Solar panels and related technologies have become more affordable and efficient over the years, making them an attractive investment for businesses. By generating their electricity, commercial buildings can reduce their reliance on the grid and lower their energy bills. Additionally, some regions offer incentives such as tax credits and grants, making solar energy an even more financially viable option.

The market for solar energy solutions is also creating new opportunities for job growth and economic development. As the demand for solar installations increases, so does the need for skilled workers. Solar panel manufacturers, installers, and maintenance technicians are just a few examples of the job opportunities emerging in this field. This growth not only benefits individuals but also stimulates local economies.

Furthermore, the market's shift towards solar energy solutions encourages innovation and technological advancements. Companies are investing in research and development to improve the efficiency and effectiveness of solar panels and related technologies. This continuous innovation drives down costs and enhances the overall performance of solar energy systems, making them more accessible to a broader audience.

In conclusion, the potential impact of solar energy on the market is immense. From reducing dependence on fossil fuels to creating new job opportunities, solar energy solutions for commercial buildings offer numerous benefits for everyone. As technology continues to evolve, solar energy will play an increasingly significant role in powering progress towards a sustainable and clean energy future.

Chapter 9: Conclusion: Embracing Solar Energy for Commercial Buildings

Importance of Renewable Energy Transition

In recent years, the urgency to transition to renewable energy sources has become undeniable. The world is facing the dual challenges of climate change and energy security, making the shift towards renewable energy an imperative for the future. Solar energy, in particular, has emerged as a leading solution to tackle these pressing issues and power progress in commercial buildings.

Renewable energy sources, such as solar power, offer a range of benefits that make them an attractive alternative to traditional fossil fuel-based energy. Firstly, solar energy is clean and sustainable, emitting zero greenhouse gases during operation. By harnessing the power of the sun, we can significantly reduce carbon emissions and combat the devastating effects of climate change.

Moreover, the transition to renewable energy sources enhances energy security. Unlike fossil fuels, which are finite and subject to price volatility, solar energy is abundant and freely available. By investing in solar panels and other solar energy solutions, commercial buildings can reduce their reliance on the grid and mitigate the risks associated with energy price fluctuations. This transition empowers businesses to take control of their energy consumption and costs, promoting long-term stability and resilience.

The importance of the renewable energy transition extends beyond environmental and economic benefits. It also creates new opportunities for job creation and economic growth. The solar energy industry has experienced significant growth in recent years, with increasing demand for solar panels and installation services. This growth translates into job opportunities in manufacturing, installation, maintenance, and research and development. By embracing solar energy solutions, commercial buildings can contribute to the growth of a green economy and support the creation of sustainable employment opportunities.

Furthermore, transitioning to solar energy can enhance a company's reputation and brand image. In an era where sustainability is at the forefront of consumer and investor preferences, businesses that demonstrate a commitment to renewable energy are regarded more favorably. Going solar not only showcases environmental responsibility but also positions companies as leaders in innovation and forward-thinking.

In conclusion, the importance of renewable energy transition, particularly in the niche of solar energy, cannot be overstated. It offers a multitude of benefits, including environmental protection, energy security, economic growth, and enhanced brand image. By embracing solar energy solutions, commercial buildings can play a pivotal role in powering progress towards a sustainable and prosperous future for everyone.

Long-Term Sustainability

In today's rapidly evolving world, the need for sustainable energy solutions has become more urgent than ever before. As we continue to witness the detrimental effects of climate change, it has become evident that we must transition towards cleaner and greener alternatives. Solar energy stands out as one of the most promising solutions, offering a myriad of benefits for both the environment and the economy.

Solar energy is derived from the sun's rays, which are captured and converted into electricity through photovoltaic panels. Unlike fossil fuels, solar power is a renewable energy source that does not deplete natural resources or emit harmful greenhouse gases. This makes it a crucial tool in combating climate change and reducing our carbon footprint. By utilizing solar energy, we can significantly reduce our dependence on non-renewable resources and contribute to a cleaner and healthier planet for future generations.

Commercial buildings play a pivotal role in the adoption of solar energy solutions. These establishments consume a substantial amount of energy, making them ideal candidates for solar power integration. By harnessing the sun's energy, businesses can not only reduce their electricity bills but also enhance their reputation as environmentally conscious entities. Embracing solar energy demonstrates a commitment to sustainability, which resonates positively with customers, investors, and the wider community.

Moreover, solar energy solutions offer long-term financial benefits. While the initial investment may seem significant, the long-term savings on energy costs can outweigh the upfront expenses. Solar panels have a lifespan of 25-30 years, ensuring a consistent and reliable source of electricity. Additionally, many governments and local authorities offer incentives and tax breaks for businesses that adopt renewable energy systems, further bolstering the financial appeal of solar energy.

The technology behind solar energy is constantly evolving, making it more efficient and accessible. Advancements in photovoltaic technology have boosted the efficiency of solar panels, allowing them to generate more electricity from the same amount of sunlight. Additionally, the decreasing costs of solar panels and installation have made solar energy systems more affordable for businesses of all sizes.

In conclusion, long-term sustainability is a crucial aspect of solar energy solutions for commercial buildings. By embracing solar power, businesses can not only reduce their environmental impact but also benefit financially in the long run. With the ongoing advancements in technology and increasing accessibility, solar energy is a viable and attractive option for every business seeking to contribute to a greener future. Let us all join hands in powering progress through solar energy and collectively drive positive change for a sustainable tomorrow.

Economic Benefits

Solar energy has emerged as a game-changer in the commercial sector, offering numerous economic benefits for businesses of all sizes. In this subchapter, we will delve into the various ways in which solar energy solutions can power progress and boost financial well-being for commercial buildings.

One of the primary economic advantages of solar energy is cost savings. By harnessing the power of the sun, businesses can significantly reduce their electricity bills. The abundance of solar energy means that once the initial investment is made, the cost of generating electricity becomes nearly free. This translates into substantial savings over time, allowing businesses to allocate their financial resources to other critical areas.

Moreover, many governments and local authorities provide attractive incentives and tax benefits for businesses that adopt solar energy solutions. These incentives can include grants, tax credits, and favorable financing options, further lowering the overall cost of installing solar systems. By taking advantage of these incentives, businesses can accelerate their return on investment and enhance their bottom line.

Solar energy also offers an opportunity for businesses to generate additional revenue streams. Through net metering programs, excess electricity produced by commercial buildings can be fed back into the grid, allowing businesses to earn credits or even sell the surplus energy to utility companies. This not only offsets the initial investment but can also create a steady stream of income for businesses in the long run.

Furthermore, solar energy solutions can enhance the value of commercial buildings. Studies have shown that properties equipped with solar systems have higher resale values and attract a larger pool of potential buyers or tenants. This increased property value can be a significant advantage for businesses looking to sell or lease their commercial spaces in the future.

Lastly, adopting solar energy solutions demonstrates a commitment to sustainability and corporate social responsibility. As more consumers prioritize environmentally-friendly practices, businesses that embrace solar energy can attract a wider customer base, resulting in increased brand loyalty and market share.

In conclusion, solar energy provides a myriad of economic benefits for commercial buildings. From cost savings and government incentives to additional revenue streams and increased property value, solar energy solutions offer a compelling proposition for businesses of all kinds. By embracing solar power, companies can not only power progress but also bolster their financial well-being in an ever-evolving economic landscape.

Environmental Stewardship

In today's world, the importance of environmental stewardship cannot be overstated. As we face the challenges of climate change and dwindling natural resources, it is crucial that we take responsibility for our actions and make conscious efforts to protect our planet. One of the most effective ways to do this is through the use of solar energy solutions in commercial buildings.

Solar energy is a renewable and sustainable source of power that harnesses the sun's rays to generate electricity. By utilizing solar energy systems, commercial buildings can significantly reduce their carbon footprint and contribute to a cleaner and greener future. This is particularly important in the context of commercial buildings, as they are known to be significant energy consumers and contributors to greenhouse gas emissions.

Solar energy solutions offer numerous benefits for both the environment and businesses. Firstly, they provide a clean and renewable source of energy that does not produce harmful emissions or pollutants. This means that by switching to solar power, commercial buildings can significantly reduce their reliance on traditional fossil fuels, such as coal or natural gas, which are major contributors to air pollution and global warming.

Furthermore, solar energy solutions can also help businesses save money in the long run. Although the initial investment in installing solar panels may seem daunting, the long-term cost savings are undeniable. By generating their own electricity, commercial buildings can reduce their dependence on the grid and lower their energy bills. Additionally, many governments and utility companies offer incentives and tax credits for businesses that invest in renewable energy, further enhancing the financial benefits of solar energy solutions.

Moreover, solar energy solutions can enhance a company's reputation and brand image. In today's increasingly eco-conscious society, consumers are actively seeking out businesses that demonstrate a commitment to sustainability and environmental stewardship. By adopting solar energy solutions, commercial buildings can showcase their dedication to reducing their environmental impact and attract environmentally conscious customers.

In conclusion, environmental stewardship is a vital concept that should be embraced by all. Solar energy solutions offer a powerful way for commercial buildings to contribute to a cleaner and more sustainable future. By harnessing the sun's energy, these buildings can drastically reduce their carbon footprint, save money, and enhance their brand image. It is time for every business to consider the adoption of solar energy solutions and become a part of the movement towards a greener world.

Steps to Getting Started with Solar Energy

Solar energy is a rapidly growing field that offers numerous benefits for both individuals and the environment. By harnessing the power of the sun, we can reduce our reliance on fossil fuels, decrease our carbon footprint, and save money on energy bills. If you are interested in exploring solar energy for your home or business, here are some essential steps to help you get started.

1. Assess your energy needs: Before diving into solar energy, it is crucial to evaluate your current energy consumption. Determine how much electricity you use on average and identify areas where you can make energy-efficient changes. This will help you determine the size of the solar system you need and the potential savings you can achieve.

2. Conduct a site analysis: Assess your property's solar potential by examining the available space for solar panels. Consider the orientation, shading, and structural integrity of your roof, as well as any local regulations or restrictions. A professional solar installer can help you with this analysis and provide recommendations for the most optimal solar system design.

3. Research available incentives and financing options: Solar energy is more accessible than ever, thanks to various financial incentives and financing options. Look into federal, state, and local incentives, such as tax credits, rebates, and grants, which can significantly reduce the upfront costs. Additionally, explore financing options like solar leases, power purchase agreements (PPAs), or loans that can make solar installations more affordable.

4. Choose a reputable solar installer: Selecting the right solar installer is crucial for a successful solar project. Research different companies,

read customer reviews, and evaluate their experience and certifications. Request quotes from multiple installers and compare their offers to ensure you get the best value for your investment. Remember, a reputable installer will provide a thorough assessment, offer quality equipment, and handle all necessary permits and paperwork.

5. Install and monitor your solar system: Once you have chosen an installer, they will handle the installation process, including mounting the solar panels, connecting them to the electrical grid, and setting up any necessary monitoring systems. Regularly monitor your system's performance to ensure optimal efficiency and address any issues promptly. Many solar systems come with online monitoring platforms that allow you to track your energy production and consumption in real-time.

By following these steps, you can embark on your solar energy journey with confidence. Solar energy not only benefits the environment but also offers long-term financial savings. Start exploring solar options in your area today and join the growing movement towards a cleaner, more sustainable future.

Assessing Feasibility

In the quest for sustainable and clean energy sources, solar power has emerged as a leading contender. With its abundant availability and the potential to reduce greenhouse gas emissions, solar energy offers a promising solution for commercial buildings. However, before embarking on a solar energy project, it is crucial to assess its feasibility.

Feasibility assessment is an essential step in determining whether solar energy solutions are viable for commercial buildings. This process involves evaluating various factors such as location, building structure, energy requirements, and financial considerations. By conducting a thorough assessment, businesses can make informed decisions about implementing solar energy systems.

One of the primary considerations in assessing feasibility is the location of the commercial building. Solar energy relies on sunlight, making it essential to determine the amount of solar irradiation received at the site. Factors such as shading from nearby buildings or trees, as well as the orientation and tilt of the roof, must be evaluated to ensure optimal solar exposure. Additionally, the local climate and weather patterns play a role in determining the feasibility of solar energy generation.

Another crucial aspect is the building structure itself. Assessing the structural integrity of the roof and determining its load-bearing capacity is essential to gauge the suitability for installing solar panels. Moreover, the availability of suitable space for mounting the panels and connecting them to the electrical system should also be considered.

In addition to physical considerations, energy requirements are a vital component in assessing feasibility. Understanding the building's current and future energy consumption helps determine the size and capacity of the solar energy system needed. By analyzing energy usage patterns and identifying opportunities for energy efficiency improvements, businesses can optimize their solar energy solutions.

Financial considerations are significant when assessing the feasibility of solar energy solutions. While the cost of solar panels has decreased over the years, it is crucial to evaluate the return on investment and payback period. Assessing available incentives, such as tax credits or grants, can also help determine the financial viability of solar energy projects.

To sum up, assessing the feasibility of solar energy solutions for commercial buildings is a crucial step before implementation. By considering factors such as location, building structure, energy requirements, and financial aspects, businesses can make informed decisions about harnessing the power of the sun. With careful assessment, solar energy has the potential to power progress and pave the way for a sustainable future for everyone.

Finding Qualified Installers

One of the most crucial aspects of implementing solar energy solutions for commercial buildings is finding qualified installers. The success of any solar project heavily relies on the expertise and experience of the installation team. Therefore, it is essential to carefully select qualified professionals who can ensure the efficient and effective installation of solar energy systems.

When embarking on a solar energy project, it is recommended to engage in thorough research to identify reputable and experienced installers in your area. Start by seeking references and recommendations from trusted sources such as colleagues, friends, or industry associations. These recommendations can provide valuable insights into the installer's reputation and the quality of their work.

Additionally, it is essential to verify the installer's qualifications and certifications. Look for installers who are certified by recognized organizations such as the North American Board of Certified Energy Practitioners (NABCEP). NABCEP certification ensures that the installer has met rigorous standards in terms of knowledge and expertise in solar energy installation.

Furthermore, it is crucial to evaluate the installer's previous work and portfolio. Review the projects they have completed in the past, paying attention to the scale and complexity of the installations. This will give you an idea of their capabilities and whether they are suitable for your project. Additionally, consider reaching out to their previous clients to gather feedback on their experience working with the installer.

Another aspect to consider when selecting installers is their familiarity with local regulations and permitting processes. Solar energy installations often require various permits and compliance with local building codes. Therefore, it is advantageous to choose installers who have experience working in your region and are well-versed in the local regulations.

Lastly, it is advisable to request multiple quotes from different installers to compare pricing and services. However, it is essential not to base your decision solely on cost. While budget considerations are important, it is crucial to prioritize quality and expertise to ensure a successful solar energy project.

In conclusion, finding qualified installers is a critical step in implementing solar energy solutions for commercial buildings. Thorough research, references, certifications, portfolio evaluation, and familiarity with local regulations are essential factors to consider when selecting installers. By choosing experienced professionals, you can ensure a seamless and successful installation of solar energy systems, driving progress towards a sustainable and environmentally-friendly future.

Maximizing Financial and Environmental Returns

In today's fast-paced world, the need for sustainable and clean energy solutions has become more urgent than ever. As we strive towards a greener future, solar energy has emerged as a powerful ally in our fight against climate change. This subchapter aims to explore how solar energy can not only help us protect the environment but also maximize financial returns for commercial buildings.

Solar energy, derived from the sun's abundant and renewable rays, offers an array of benefits that extend beyond reducing carbon emissions. One of the most compelling advantages is the significant cost savings it brings to commercial buildings. By harnessing the power of the sun, businesses can drastically reduce their dependence on traditional energy sources, leading to substantial reductions in electricity bills. Moreover, with advancements in solar panel technology, the initial investment required for installing solar panels has considerably decreased, making it an increasingly viable and attractive option for commercial buildings of all sizes.

The financial benefits of solar energy extend beyond immediate savings on electricity bills. Many governments and jurisdictions offer attractive incentives, such as tax credits, grants, and rebates, to encourage the adoption of solar energy. These incentives can help offset the initial installation costs, further bolstering the financial returns for commercial buildings. Additionally, solar energy systems have long lifespans, requiring minimal maintenance, which translates into long-term cost savings and increased profitability for businesses.

While financial gains are undoubtedly compelling, the environmental benefits of solar energy are equally important. Solar power generates electricity without harmful greenhouse gas emissions, reducing our carbon footprint and mitigating the adverse effects of climate change. By embracing solar energy solutions, commercial buildings can play a pivotal role in transitioning towards a cleaner and more sustainable energy future.

Moreover, adopting solar energy can enhance a company's reputation and brand image. In an era where consumers are increasingly conscious of environmental issues, businesses that prioritize sustainability and renewable energy are more likely to attract eco-minded customers and investors. By investing in solar energy, commercial buildings can position themselves as leaders in their industry, setting an example for others to follow.

In conclusion, maximizing financial and environmental returns through solar energy solutions is a win-win situation for commercial buildings. By embracing solar power, businesses can reduce operating costs, take advantage of government incentives, and enhance their environmental stewardship. The time to harness the power of the sun is now, and commercial buildings have a unique opportunity to lead the charge towards a brighter and more sustainable future.